Forest and Plant Bioproducts Division 2013

Core Programming Area at the 2013 AIChE Annual Meeting: Global Challenges for Engineering a Sustainable Future

San Francisco, California, USA
3-8 November 2013

ISBN: 978-1-63439-042-2

Printed from e-media with permission by:

Curran Associates, Inc.
57 Morehouse Lane
Red Hook, NY 12571

Some format issues inherent in the e-media version may also appear in this print version.

Printed by Curran Associates, Inc. (2014)

For permission requests, please contact AIChE
at the address below.

AIChE
3 Park Avenue
New York, NY 10016-5991

Phone: (203) 702-7660
Fax: (203) 775-5177

www.aiche.org

Additional copies of this publication are available from:

Curran Associates, Inc.
57 Morehouse Lane
Red Hook, NY 12571 USA
Phone: 845-758-0400
Fax: 845-758-2634
Email: curran@proceedings.com
Web: www.proceedings.com

TABLE OF CONTENTS

(82a) A Novel Fast Pyrolysis Model Based On Variable-Power Image Furnace Results

Mauviel, G., LRGP, ENSIC, CNRS-Nancy Universite
Christodoulou, M.,
Lede, J., LRGP, CNRS - Universite de Lorraine
Billaud, F.,
Originally presented on: 11/4/2013 8:30:00 - 8:52:00

A Novel Fast Pyrolysis Model Based On Variable-Power Image Furnace Results

Guillain MAUVIEL, LRGP, CNRS - Universite de Lorraine

Melina CHRISTODOULOU, LRGP, CNRS - Universite de Lorraine

Jacques LÉDÉ, LRGP, CNRS - Universite de Lorraine

Francis BILLAUD, LRGP, CNRS - Universite de Lorraine

1 rue Grandville, 54001 Nancy Cedex, France

T: +33(0)3.83.17.52.07, guillain.mauviel@univ-lorraine.fr

Nowadays, ligno-cellulosic biomass is considered as a renewable carbon resource. It represents a relevant alternative to reduce greenhouse gases emissions. During last decades, many scientists and industrials have been interested in thermochemical conversion of biomass such as gasification. In order to accurately model the gasifiers, it is important to study the implied thermochemical phenomena (such as pyrolysis), through dedicated experiments close to the industrial conditions. This communication presents the model developed by LRGP to describe primary pyrolysis in bubbling fluidized beds conditions, one of the major technologies used for biomass gasification. The novelties of this proposed model are the coarse description of the 3 main products (gas, vapours, char) and the conservation of the 3 main biomass elements: carbon, hydrogen, oxygen. The work is part of the Gaya project, led by GDF Suez and funded by ADEME. Gaya project will contribute to develop a new gasification demonstration platform in Europe for SNG production.

The experimental setup used to adjust the pyrolysis model is a vertical image furnace [Christodoulou & al 2012]. The principle is to concentrate the radiation of a xenon arc lamp on a biomass sample. The lamp is settled at the first focus of an elliptical mirror. The light beams are concentrated at the second focus where the biomass sample is placed inside a transparent reactor. In order to prevent any gas-phase secondary reactions, the volatile matter (vapours and gases) released during the reaction is quenched by a controlled flow of nitrogen at room temperature. During a flash time, the temperature evolution is measured by pyrometry on both sides. As soon as the light beams are switched off, all the pyrolysis products can be recovered, allowing achieving complete mass balances. Gases are collected in a sampling bag, liquids are condensed and aerosols are trapped in a filter. The solid residue remains on the sample holder in the reactor. All these products are analysed off-line. The heat flux density applied to the biomass can reach 650 kW m^{-2}. This is the maximum value encountered by wood particles surrounded by hot bed particles in fluidized bed gasifier. The heat flux density is decreased over time in order to reproduce the thermal story of a wood particle injected inside a fluidized bed.

The single biomass particle pyrolysis model which has been developed is one dimensional and based on a classical simple reaction scheme: three competitive first-order reactions [Chan & al 1985, Al-Haddad & al 2010]. The model numerical resolution allows computing the three main products yields and the temperature gradient inside a wood chip. The kinetic and stoichiometric parameters of this model are fitted thank to the experimental data obtained with the image furnace (temporal evolution of gas / char / water / organic vapour yields, gas composition, solid residue ultimate analysis). The interest and originality of this model is to rely on the establishment of atomic balances for all products. It permits to describe the gas molecular composition, a global atomic composition of tar and a spatial (along the thickness) atomic composition of the char. The comparison between the experimental data and simulated results for a cylindrical pellet will be shown. The extrapolation capacity of this model will be assessed.

References

M. Christodoulou, G. Mauviel, J. Lede, P. Beaurain, M. Weber, H.Legall, F. Billaud, J. Anal. Appl. Pyrol., 2012, DOI: 10.1016/j.jaap.2012.11.006.

M. Al-Haddad, E. Rendek, J.-P. Corriou, G. Mauviel, Energ. Energy Fuels., 2010, 24, 4689.

W. C. R Chan, M. Kelbon, B. B. Krieger, Fuel, 1985, 64, 1505.

(82b) Modeling Biomass-to-Fuels Processing

Moreno, B. M., University of Delaware
Klein, M. T., University of Delaware
Originally presented on: 11/4/2013 8:52:00 - 9:14:00

Fuels produced from lignocellulosic biomass are a promising renewable energy source that will aid in the transition from petroleum-based liquid fuels for transportation applications. Lignocellulosic biomass is composed of lignin, cellulose, and hemicellulose, in varying amounts for each plant species. Pyrolysis of these biopolymers results in the production of aromatic compounds, alkanes, and various oxygenated species.

The high oxygen content of the biomass pyrolysis oil must be decreased for use in conventional internal combustion engines. Hydrodeoxygenation of biomass pyrolysis oil results in a mixture of hydrocarbons with chemical properties similar to those of petroleum-derived fuels. Co-processing of hydrotreated biomass pyrolysis oil in a traditional petroleum refinery provides a transitional renewable energy source for liquid transportation fuel applications while utilizing the existing processing and distribution infrastructure. Production of biofuels involves novel molecular species and unknown chemical reaction pathways that are well-suited for kinetic modeling.

Molecular-level modeling of the thermochemical conversion of biomass into renewable fuels requires the creation of a complete reaction network that identifies each chemical pathway from the lignocellulosic biomass feedstock to the desired fuel products. The Klein research group has developed a set of software tools that are capable of rapidly generating the complete reaction network and associated kinetic model for this and other applications. A robust reaction network is first produced by INGen (the Interactive Network Generator) [1] from the molecular starting materials. KME (the Kinetic Modeling Editor) [2,3] automatically writes the corresponding rate equations and mass balances for each molecular species. The resulting molecular-level kinetic model is subsequently numerically solved to provide model predictions of the product composition at varying process conditions.

This biomass-to-fuels processing model provides insight into the competing kinetic pathways for the production of renewable liquid fuels from lignocellulosic materials. Predictions are made for product composition at varying feedstock composition and processing conditions. Comparison of the activation energy barriers for parallel deoxygenation routes will direct catalyst research to modulate between these pathways and provide the highest quality products at the mildest process conditions. The ability to rapidly generate and solve molecular-level kinetic models will enable real-time optimization of fuel production and increase the viability of biofuels in the renewable energy market.

[1] Bennett, C.A. PhD Dissertation, Chemical & Biochemical Engineering, Rutgers University. 2009.

[2] Hou, Z.; Bennett, C.A.; Klein M.T.; Virk, P.S. *Energy & Fuels*. 2010. *24* (1), 58-67.

[3] Wei, W.; Bennett, C.A.; Tanaka, R.; Hou, G.; Klein, M.T. Jr.; Klein, M.T. *Fuel Processing Technology*. 2008. *89* (4), 350-363.

(82c) Global Kinetics for the Simulation of Biomass Pyrolysis in Fluidized Bed Reactors

Pepiot, P., Cornell University
Malhotra, K., Cornell University
Ciesielski, P. N., Vanderbilt University
Grout, R., National Renewable Energy Laboratory
Nimlos, M., National Renewable Energy Laboratory
Originally presented on: 11/4/2013 9:14:00 - 9:36:00

Accounting adequately for the chemical processes and their interactions with the surrounding flow dynamics is an essential step toward predictive Computational Fluid Dynamic (CFD) simulations of biomass-to-biofuel thermochemical conversion. The direct implementation of detailed kinetic mechanisms, involving a large number of chemical species and elementary reactions, is usually proscribed due to limited available computational resources. While numerous chemistry reduction techniques have been developed to generate reduced-order kinetic models from more detailed kinetic schemes, these techniques are unable to handle the complex, multiphasic nature of biomass chemistry, and in general, cannot achieve the very high level of reduction required by most CFD tools. A systematic strategy is presented here to reduce a given detailed kinetic model into a global model that will: (i) contain many fewer reaction steps, (ii) use lumped variables that combine species of similar chemical nature, and (iii) maintain the predictive capabilities of the detailed mechanism for the quantities of interest. The approach is demonstrated in the context of biomass pyrolysis in fluidized bed reactors. Partially stirred reactor (PaSR) models are used in conjunction with existing detailed chemical kinetic mechanisms for both solid and gas phase chemistry, to generate a database of compositions likely to occur in an actual reactor. From the analysis of this database, representative lumped species, and an appropriate set of global reactions and corresponding kinetic rates are identified automatically. Validation is performed in various configurations, and the resulting global model is integrated into a Lagrangian particle tracking tool recently developed for the large-scale simulation of reactive granular flows.

(82d) An Open-Source Code for Simulating Biomass Pyrolysis in Fluidized-Bed Reactors

Xiong, Q., Iowa State University
Kong, S. C., Iowa State University
Originally presented on: 11/4/2013 9:36:00 - 9:58:00

A numerical framework named BIOTC (BIOmass Thermochemical Conversion) for simulating biomass fast pyrolysis is developed in this study. Fast pyrolysis, a thermochemical conversion process converting low-value lingo-cellulosic biomass to various useful products, has gained increased interest. This study is to develop an open-source computational tool to help understand the complex phenomena involved within the biomass fast pyrolysis process. In BIOTC, a multi-fluid model is applied to simulate the multi-phase hydrodynamics while global reaction kinetics is used to describe the physicochemical conversion. The coupling of these two methodologies is realized by a time-splitting method. The model results are validated using experimental data from fast pyrolysis reactors. Good levels of agreement are obtained in the product distribution including tar, biochar, and syngas. The parametric study also indicates that the tar yield can be increased if the biomass injector location is elevated or the superficial velocity of the feeding nitrogen is increased. One characteristics of BIOTC is that sub-models for different constitutive hydrodynamic relations and chemical reactions can be readily incorporated into this framework owing to the object-oriented feature of the baseline code.

(82e) Development of a Plug-Flow Reactor for Direct Hydrothermal Liquefaction of Algae and Livestock Waste

Minarick, M. J., University of Illinois at Urbana-Champaign
Originally presented on: 11/4/2013 9:58:00 - 10:20:00

Hydrothermal processing of biomass is a technology that has shown promising results in treating waste and producing oil. A batch hydrothermal processing system that was previously developed at the University of Illinois at Urbana-Champaign was used to treat swine manure and it successfully converted up to 70% of swine manure volatile solids into oil. A continuous stirred-tank reactor (CSTR) demonstrated results similar to the batch reactor and proved the feasibility of a continuous process, but CSTRs are not conducive for scale-up at the high pressures associated with hydrothermal liquefaction of biomass so a continuous plug-flow reactor was designed and implemented. After the livestock waste is converted to oil, the nutrients remaining in the waste water can be used to grow algae, which clean the wastewater and can also be hydrothermally processed into oil. The oil conversion rates associated with processing livestock waste and algae using a plug-flow reactor were compared with those from the batch reactor and the CSTR to assess the feasibility of scale-up.

(82f) Depolymerization of Lignin During Solvent Liquefaction

Kim, K. H., Iowa State university
Kieffer, M., Iowa State University
Bai, X., Iowa State University
Brown, R., Iowa State University
Originally presented on: 11/4/2013 10:20:00 - 10:42:00

Lignin, one of the main components of lignocellulosic biomass, is a complex polymer made up of monolignol monomers that are methoxylated to various degrees. . Lignin has potential as feedstock for biobased chemicals and fuels if it can be effectively depolymerized and/or deoxygenated. Solvent liquefaction is able to depolymerize lignin under relatively mild conditions and solvent combinations. However, depolymerization of lignin is usually accompanied by repolymerization of depolymerization products. We are investigating whether polymerization can be reduced or prevented under carefully controlled solvent liquefaction conditions. The effect of different solvents and solvent combinations on thermal depolymerization of lignin was investigated in micro- and gram-scale bath reactors. Organosolv lignin extracted from corn stover was reacted with various solvents in the micro reactor at temperatures from 300 to 400 °C and reaction times from 5 to 20 min. At the end of reaction volatiles released from the reactor were directly analyzed by online GC/MS. Lignin and selected solvents were also reacted in a gram-batch reactor to recover bio-oil. The molecular weight distribution of bio-oil was analyzed using GPC and bio-oil was subjected to aging tests to evaluate stability of bio-oil produced. The tested solvents were tetralin, formic acid, isopropanol, 1, 4-dioxane and water.

The preliminary results showed that hydrogen donor solvents mainly increased the yields of alkylphenolic monomers by saturating C=C bonds in the phenolic side chains and the selectivity of alkylphenols increased with increasing reaction temperature. The yield of alkylphenols first increased then decreased with longer reaction times. Secondary reactions between decomposition products were also investigated.

(82g) What Can We Learn About Fast Pyrolysis of Lignin From Solvent Liquefaction?

Brown, R., Iowa State University
Bai, X., Iowa State University
Kim, K. H., Iowa State university
Kieffer, M., Iowa State University
Originally presented on: 11/4/2013 10:42:00 - 11:04:00

Fast pyrolysis of lignocellulosic biomass produces bio-oil consisting of an emulsion of lignin-derived phenolic oligomers in an aqueous phase derived mainly from polysaccharides. The phenolic oligomers present several problems to the commercial utilization of bio-oil, including high viscosity, low volatility, and poor thermal stability. Ideally, lignin would depolymerize into more tractable phenolic monomers and dimers. Achieving such an outcome requires a better understanding of lignin pyrolysis than currently exists. Many researchers suggest than the oligomers represent fragments of lignin that have been thermally ejected from the pyrolysis zone as liquid droplets. Others suggest that these oligomers derived from volatile depolymerization products that recombine in the gas phase and nucleate to liquid aerosols.

Solvent liquefaction is an alternative thermal depolymerization process that can be described as "pyrolysis in a solvent." Investigating the chemical and physical phenomena that control thermal depolymerization of lignin in some respects is more easily conducted under conditions of solvent liquefaction than fast pyrolysis because reactions occur more slowly and in a single phase. Solvent liquefaction studies suggests the rapid and complete depolymerization of lignin to monomers and dimers followed by repolymerization to phenolic oligomers unless reaction is quenched by either dispersing or "capping" the depolymerization products. This suggests a path forward to produce more desirable fast pyrolysis products from lignin.

(145a) Structural Understanding of Key Physical Processes in the Pretreatment and Enzyme Hydrolysis of Biomass

Davison, B. H., BioEnergy Science Center and Oak Ridge National Laboratory
Originally presented on: 11/4/2013 12:30:00 - 12:55:00

Lignocellulosic biomass is a complex biological composite material that shows significant recalcitrance towards the structural deconstruction and enzymatic hydrolysis into sugars that is necessary for fermentation to bioethanol. This talk will summarize developments in the "Dynamic Visualization of Lignocellulose Degradation by Integration of Neutron Scattering Imaging and Computer Simulation" for multiple-length scale, real-time imaging of biomass during pretreatment and enzymatic hydrolysis. By combining multiple probes of structure, sensitive to different length scales, with molecular dynamics simulations, we reveal the fundamental physical processes responsible for the morphological changes in biomass during steam explosion pretreatment. We show that the basic driving forces are the same in other leading pretreatments, such as dilute acid pretreatment. The compensation between the entropy and enthalpy of hydration drives the cell wall components over kinetic barriers, destabilizing them, and causing changes that render cellulose more accessible to hydrolytic enzymes. We also utilized neutron scattering to probe cellulase enzyme structure while hydrolyzing deuterated bacterial cellulose. These show new insights into structural changes of lignin and cellulose during dilute acid pretreatment and cellulose-cellulase interactions.

(145b) The Remarkable Role of Extractable Lignin in Enhancing Enzymatic Hydrolysis of Organosolv Pretreated Biomass

Tu, M., Auburn University
Li, M., Auburn University
Lai, C.,
Originally presented on: 11/4/2013 12:55:00 - 13:20:00

The interactions between lignin and cellulase enzymes play a critical role in the effective enzymatic hydrolysis of lignocellulosic biomass. Organosolv pretreatment has shown great potential for producing lignocellulosic substrates from softwood and hardwood with good digestibility. The objective of this work was to investigate the role of extractable lignin in the enzymatic hydrolysis of ethanol organosolv pretreated sweetgum (OPSG). We removed the extractable lignin in OPSG by ethanol washing and 1% NaOH, then compared the enzymatic hydrolysis performance. Interestingly, we found that removing extractable lignin in OPSG showed an adverse effect on enzymatic hydrolysis. The hydrolysis yield dropped from 70.6% (washing only with water) to 55.5% (washing with 95% ethanol). Removing extractable lignin in OPSG with 1% NaOH also reduced the final hydrolysis yield from 69.9% to 43.8%. Furthermore, when we added extractable lignin into the hydrolysis of Avicel, and the hydrolysis yield increased from 67.6% to 71.7%. It indicated extractable lignin in pretreated substrates potentially facilitated the enzymatic hydrolysis. However, the bulk lignin still limited the enzymatic hydrolysis significantly; removing bulk lignin increased the hydrolysis yield by 49%. Therefore, we believe that extractable lignin and bulk lignin probably play remarkably different roles in enzymatic hydrolysis, and extractable lignin can assist the enzymatic hydrolysis.

(145c) Linking Glycan Accessibility to Cellulolytic Enzymes to Cell Wall Porosity and Swelling Capacity in Diverse Pretreated Biomass

Williams, D. L., Michigan State University
Hodge, D., Michigan State University
Originally presented on: 11/4/2013 13:20:00 - 13:45:00

The purpose of this work is to link the surface and pore properties of nano-scale pores within plant cell walls to enzymatic hydrolysis, enzyme adsorption, and enzyme accessibility through an understanding of the properties impacting swelling and cell wall porosity, and understand how these properties evolve during pretreatments and enzymatic hydrolysis. This involves the role of surface properties in influencing: **(1)** water infiltration into the cell wall and cell wall swelling, **(2)** cellulolytic enzyme accessibility to binding sites within the cell wall, and **(3)** enzyme binding to chemically and physically modified surfaces within the cell wall. For this, we propose to investigate the limitations of cell wall porosity and inaccessible surface area to enzymes. Specifically, using diverse biomass modified by diverse pretreatments or delignifications, we will relate composition and water properties including water swelling capacity, the water settling volume, and DSC-determined freezing point depression of cell wall-associated water within pores to cellulolytic enzyme, carbohydrate binding module (CBM), and polymer probe accessibility within the porous cell wall.

(145d) Supercritical Fluid Delignification and Enzymatic Saccharification of Agave Bagasse

Perez-Pimienta, J. A., Universidad Autonoma de Nayarit
Galicia-Luna, L. A., INSTITUTO POLITECNICO NACIONAL-ESIQIE
Elizalde-Solis, O., INSTITUTO POLITECNICO NACIONAL-ESIQIE
Stavila, V., Sandia National Laboratories
Cheng, G., Lawrence Berkeley National Laboratory
Singh, S., Joint BioEnergy Institute, Emeryville, CA 94608 and Sandia National laboratories, Livermore, CA
Simmons, B. A., Joint BioEnergy Institute, Emeryville, CA 94608 and Sandia National laboratories, Livermore, CA
Ponce-Noyola, M. T.,
Originally presented on: 11/4/2013 13:45:00 - 14:10:00

Biomass recalcitrance to saccharification is one of the major obstacles to cost-efficient production of fuels and chemicals from lignocelluloses. However, lignocellulosic biomass is recalcitrant to effective enzymatic hydrolysis because of its structural complexity, such as high degree of polymerization of cellulose, high crystallinity, presence of lignin and hemicelluloses, among others. A pretreatment operation is typically necessary to: (1) efficiently and economically break down the 3D structural network of cellulose, hemicellulose and lignin in the plant cell walls of lignocellulosic biomass, and (2) disrupt the tight hydrogen-bonded microcrystalline cellulose, both of which contribute to the recalcitrance of biomass to subsequent enzymatic saccharification. Lignin blocked cellulose accessibility to cellulose and decreases cellulose activity by competitively binding to hydrolytic enzymes. Delignification of agave bagasse using supercritical carbon dioxide with ethanol as co-solvent using different temperatures, pressures and solvent concentration was employed. Subsequently the biomass was hydrolyzed after pretreatment using cellulases combined with β–glucosidase and analyzed by X-ray diffraction, Fourier transform infrared spectroscopy, and wet chemistry methods.

(145e) Improving Glucose Yield By Fractionation of Biomass With Non-Cl? Ionic Liquids

Lynam, J. G., University of Nevada, Reno
Coronella, C. J., University of Nevada, Reno

Originally presented on: 11/4/2013 14:10:00 - 14:35:00

Lignocellulosic biomass is a sustainable, available energy resource that is mostly carbon-neutral. To convert it to liquid fuels by biochemical means, however, requires pretreatment of biomass to remove lignin to enhance the availability of cellulose. Most standard lignin removal processes use either high temperatures and pressures, hazardous acids or alkalis, or some combination of these steps.

Ionic liquids are molten salts with melting points below 100 °C, some of which have been shown to dissolve lignin. However, many such ionic liquids frequently contain Cl^- ions, which can interfere with downstream processing, e.g., enzymatic hydrolysis. Ionic liquids, with their low vapor pressures and potential for recycling, are a green alternative for lignin removal at low temperatures.

In this study, the effects of non- Cl^- containing ionic liquids 1-ethyl-3- methylimidazolium acetate (EMIM Ac), 1-ethyl-3- methylimidazolium formate (EMIM Formate), and 1-ethyl-3-methylimidazolium lactate (EMIM Lactate) were examined. The biomass studied were rice hulls, corn stover, switchgrass, loblolly pine, and poplar. The experiments used a 10% (w/w) loading of biomass in ionic liquid, a temperature of 110°C, and a fractionation time of 3 hours. The effects of biomass type and ionic liquid type were investigated by performing enzymatic hydrolysis on raw and pretreated biomass types. The non- Cl^- containing ionic liquids removed up to 98% of the lignin from biomass, which would be expected to make cellulose more available for further processing. However, EMIM lactate was less effective than EMIM Ac and EMIM Formate for each biomass type. When enzymatic hydrolysis was performed, only biomass pretreated with EMIM Ac and EMIM Formate showed greatly improved glucose yield compared to the raw biomass.

This pretreatment step using environmentally-friendly, non- Cl^- ionic liquids to remove lignin enables a green pathway from lignocellulosic biomass through enzymatic hydrolysis and fermentation to biofuels. This pretreatment step might permit more cost effective cellulosic biofuels.

(145f) A Novel Chemical Pretreatment Method to Accelerate Bioconversion Rate of Renewable Feedstocks

Kang, Y., Georgia Institute of Technology
Sohn, M. J., Georgia Institute of Technology
Hoffman, C.,
Realff, M., Georgia Institute of Technology
Bommarius, A. S., Georgia Institute of Technology
Originally presented on: 11/4/2013 14:35:00 - 15:00:00

Abstract

Lignocellulosic biomass is the most abundant naturally renewable organic resource for biofuels production. Due to its recalcitrance to hydrolysis, pretreatment is a crucial step prior to enzymatic hydrolysis of the feedstock. A variety of pretreatment methods have been intensively studied to achieve optimal yield without imposing significant adverse impact on the environment. Herein, we present a novel and cost-effective chemical pretreatment method using substituted heterocycles in an environmentally friendly way.

Methylimidazole is a precursor to some imidazolium-based ionic liquids. It can be used as a pretreatment agent at mild conditions, i.e. ambient temperature and pressure, for a very short period of time to obtain a substantial increase in the hydrolysis rate throughout the conversion of cellulose substrate. X-ray diffractometry was utilized to study the crystal structure of cellulose and a reduction in crystallinity index was observed, which can be a crucial contributor to the improvement of enzymatic degradation rate. Furthermore, recycling of methylimidazole is possible and we have tested the recyclability by reusing methylimidazole for up to four times without observing marked reduction in the initial 1-hour hydrolysis rate of cellulose. The pretreatment effectiveness of methylimidazole on both original and steam-exploded lignocellulosic biomass has been investigated. Improvement in initial hydrolysis rate was observed for both lignocellulosic substrates.

(291c) Effect of Particle Size and Temperature On Woody Biomass Slow Pyrolysis Process

Bennadji, H., Cornell University
Smith, K., Cornell University
Fisher, E., Cornell University
Originally presented on: 11/4/2013 12:30:00 - 12:55:00

Slow pyrolysis of dried pine spheres was carried out in a flow reactor to study the influence of temperature (378-465 °C) and particle size (2.54-3.81 cm) on the product distribution and temperature profiles. Pyrolysis was performed by inserting spheres into a heated nitrogen flow for a residence time of 40 min. The particle size of the samples used in the experiment is limited by the reactor diameter. The temperature histories within the wood spheres were measured at three radial locations: center (r=0), and two ?middle? locations (r= R/2), using three sheathed 0.5 mm OD K-type thermocouple inserted through holes drilled in the particle. Figure 1 shows the installation of thermocouples inside the sphere. The temporal evolution of the species (CO, CO_2, CH_4, HCHO, CH_3OH, HCOOH, and CH_3COOH) released during the sample pyrolysis was measured using Fourier Transform Infrared (FTIR) analyzer. TCD and MS detectors on a gas chromatograph were used to analyze the time-resolved concentration of hydrogen, and the tars collected over the course of the entire experiment, respectively.

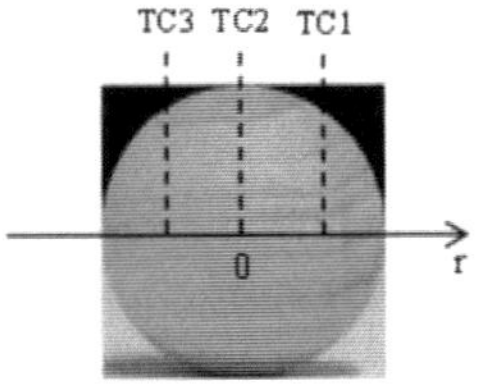

Figure 1: Thermocouples locations within the woody sphere.Â The arrow indicates the nitrogen flow direction.

Figure 2 illustrates the temperature history of the two particle sizes (2.54 cm and 3.81 cm) heated with a nitrogen flow at 466 °C. Both profiles show two thermal events; an endothermic reaction, followed by an exothermic reaction. The temperature peak appears early for the 2.54 cm sphere at 5.33 min however for 3.81 cm, it appears at 9.77 min. This trend is reasonable because for the large particle the heat transfer is slower.

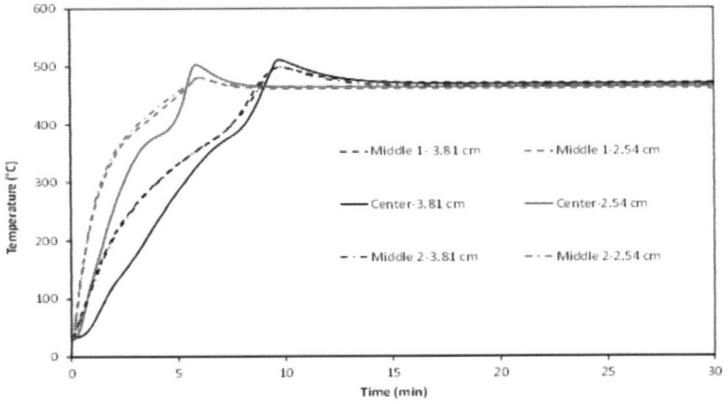

Figure 2: Particle temperature history comparison at a nitrogen temperature of 466 °C.

Figure 3 below illustrates the species time histories at a pyrolysis temperature of 466°C and for sphere 2.54 cm diameter, displaying a high degree of dilution caused by the large nitrogen flow (5.8E-03 Kg/s). Different classes of species were identified, including hydrocarbons, aldehydes, alcohols, and carboxylic acids, in addition to CO, CO_2 and H_2. Good repeatability was obtained over at least two runs. As observed in this figure, the formation of all detected species, except for methane, occurs very early (starting earlier than 1 min after sphere insertion). The peak concentrations of CO_2, HCHO, CHOOH and CH_3COOH appear earlier than the other measured species. Furthermore, the results indicated the presence of the same main species for the both particle sizes and the three temperature conditions.

Figure 4 shows the char, tar and gases yields measured for the two particle sizes and the three pyrolysis temperatures. Gases are defined as CO, CO2, and CH4 only, and tar yields in the experiment are obtained by difference.

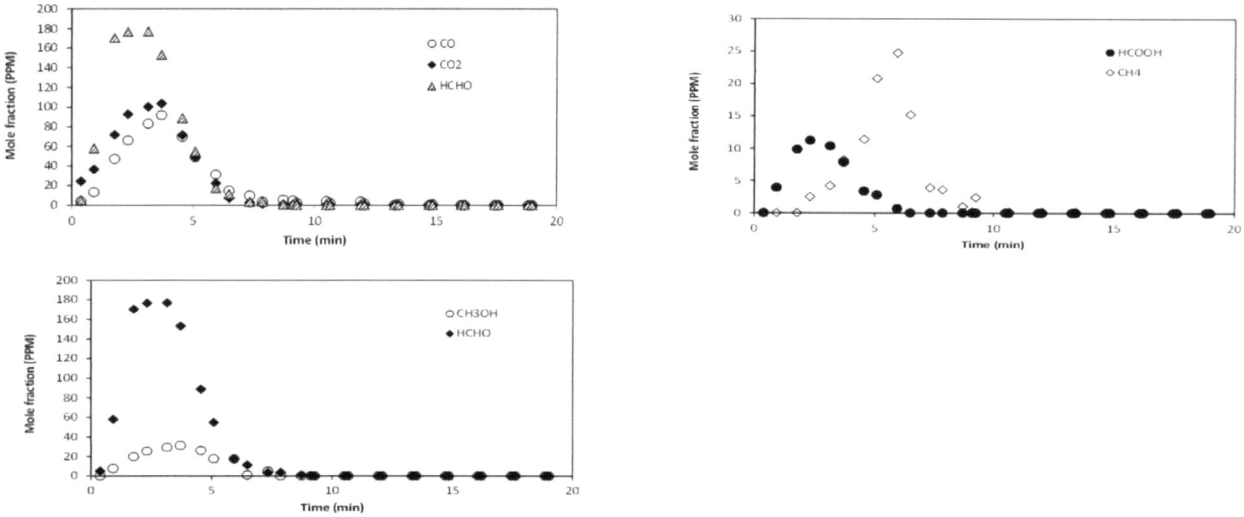

Figure 3: Mole fractions of species released from 2.54 cm pine sphere at a nitrogen temperature of 466°C.

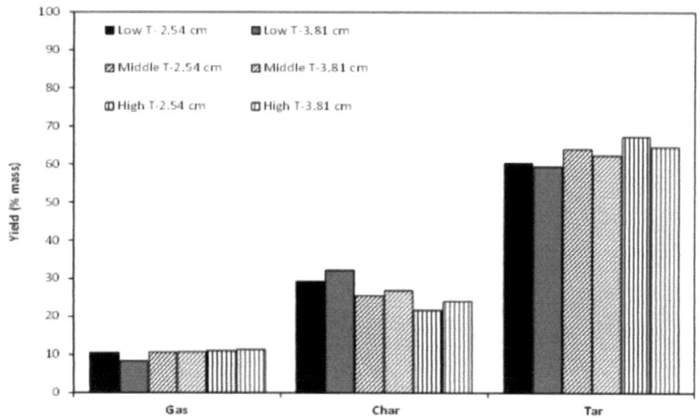

Figure 4: Experimental products yields (mass % as a percent of initial wood mass) obtained for the two particle sizes at three temperatures.

The experimental results were then compared to those calculated with a model developed by Park et al. (2010)[1]. This model was developed to determine the kinetics involved in the slow pyrolysis of large wood particles under conditions of low temperature and long residence times. It includes transient mass, species, and energy conservation equations, in addition to the Darcy momentum equation. Coupled to these equations is the multistep kinetic mechanism describing direction production of gas and tar, and indirect production of char. Each reaction is assumed to be first order with Arrhenius type rate expression.

Because the pyrolysis model was developed for a specific experimental configuration, changes were made to better represent the results obtained from the experimental setup presented here. Modifications included an adaptation of some wood and char properties, such as thermal diffusivity, and an experimental determination of thermal boundary conditions. The modified pyrolysis model was implemented in COMSOL, a commercial Multiphysics software.

[1] W.C. Park, A. Atreya and H.R. Baum, Combustion and Flame 157 (2010) 481-494

(147a) Effect of Various Species of Woody Biomass On Bio-Oil Properties and Its Correlations

Ren, X., Beijing Forestry University
Meng, J., North Carolina State University
Moore, A., North Carolina State University
Chang, J., Beijing Forestry University
Park, S., North Carolina state university
Originally presented on: 11/4/2013 12:55:00 - 13:20:00

Lignocellulosic biomass has received significant attention due to the limited reserve of petroleum and the increasing energy consumption. Woody biomass is an ideal renewable resource in terms of high bulk density, steady constituents, and great amounts of reserves. Fast pyrolysis has shown a great potential to convert biomass into intermediate bio-oil for transportation fuels and chemicals. The physicochemical properties of biomass are important for thermochemical process, affecting both the process efficiency and product properties. However, the detailed relationship is not fully understood. Moreover, there are few reports on the prediction of bio-oil properties based on feedstock characteristics. This study aims to extensively explore the correlations among feedstock characteristics (e.g. proximate and ultimate analysis, inorganic content and composition, chemical composition analysis) with thermochemical reactivity (e.g. bio-oil yield, pyrolytic activation energy) and bio-oil properties (e.g. oxygen content, stability, chemical composition). Several hardwoods (eucalyptus, sweet gum, beech, sourwood, red maple, acacia), hardwood barks, softwood (loblolly pine), and non-woody biomass (switchgrass, miscanthus) were used in this study. Based on the sequential data of "biomass-pyrolysis-bio-oil", correlations among these parameters were developed, which could be useful to better understand the process and screen the feedstocks. The results from this work are building a bridge between feedstock characteristics and its thermochemical reactivity to produce valuable bio-oil intermediate.

(147b) Reactant Conversion and Product Speciation of Carbohydrate Model-Compound Pyrolysis Using a Novel Flow Reactor

Fahey, P. J., North Carolina State University
Seshadri, V., North Carolina State University
Geng, X., North Carolina State University
Westmoreland, P. R., North Carolina State University
Originally presented on: 11/4/2013 13:20:00 - 13:45:00

Reactant Conversion and Product Speciation of Carbohydrate Model-Compound Pyrolysis Using a Novel Flow Reactor

Patrick J. Fahey, Vikram Seshadri, Xinglian Geng, and Phillip R. Westmoreland*

North Carolina State University, Raleigh, NC 27695

The model compounds 1,2-propanediol, 1,3-propanediol, 1,2,3-propanetriol, hydroxypropanone, and acetaldehyde diethyl acetal were pyrolyzed neat and as methanol solutions in a pulse-injected tube-furnace reactor. Reaction products were analyzed directly by comprehensive, two-dimensional gas chromatography time-of-flight mass spectrometry (GCxGC-TOFMS). For example, pyrolysis of 1,3-propanediol at 500°C produced 90 identified peaks.

We experimented with molecules smaller than glucose to understand better the pyrolysis reaction networks of glucose-based carbohydrates. Cellulose has a complicated pyrolysis reaction network that researchers have tried to understand through experimenting with monosaccharide, disaccharide, and small-oligosaccharide pyrolyses.[1,2] Here, using molecules smaller than glucose provided two benefits not possible with using entire monomers or dimers. First, smaller molecules isolate specific sequences of functional groups, which reduces the number of possible side reactions and products. Thus, smaller molecules simplified the chemical analysis of the reaction-product mixture and the construction of reaction pathways.[3] Second, smaller molecules allow computational-quantum-chemistry simulations to execute faster with the same level of theory. Thus, smaller molecules allow deeper analyses of the specific reaction mechanisms in their hypothesized pyrolysis reaction networks.[4]

References

1. Pushkaraj R. Patwardhhan, Justinus A. Satrio, Robert C. Brown, Brent H. Shanks, "Product distribution from fast pyrolysis of glucose-based carbohydrates," *Journal of Analytical and Applied Pyrolysis*, 86 (2009) 323-330.

2. Edward B. Sanders, Alan I. Goldsmith, Jeffrey I. Seeman, "A model that distinguishes the pyrolysis of D-glucose, D-fructose, and sucrose from that of cellulose. Application to the understanding of cigarette smoke formation," *Journal of Analytical and Applied Pyrolysis* 66 (2003) 29-50.
3. John B. Paine III, Yezdi B. Pithawalla, John D. Naworal, Charles E. Thomas Jr. "Carbohydrate pyrolysis mechanisms from isotopic labeling Part 1. The pyrolysis of glycerin: Discovery of competing fragmentation mechanisms affording acetaldehyde and formaldehyde and the implications for carbohydrate pyrolysis," *Journal of Analytical and Applied Pyrolysis* 80 (2007) 297-311.
4. Vikram Seshadri and Phillip R. Westmoreland, "Concerted reactions and mechanism of glucose pyrolysis and implications for cellulose kinetics," *Journal of Physical Chemistry A*, 116 (2012) 11997-12013.

Contact Information

Patrick J. Fahey

pjfahey@ncsu.edu

Department of Chemical & Biomolecular Engineering

North Carolina State University

Box 7905

Raleigh, NC 27695-7905

Vikram Seshadri

vseshad@ncsu.edu

Department of Chemical & Biomolecular Engineering

North Carolina State University

Box 7905

Raleigh, NC 27695-7905

Xinglian Geng

xgeng@ncsu.edu

Department of Chemical & Biomolecular Engineering

North Carolina State University

Box 7905

Raleigh, NC 27695-7905

Phillip R. Westmoreland*

prwestmo@ncsu.edu

Professor, Department of Chemical & Biomolecular Engineering

Executive Director, Institute for Computational Science and Engineering

North Carolina State University

Box 7905

Raleigh, NC 27695-7905

*To whom correspondence should be addressed.

(147c) Determination of Pyrolysis Oils Aging Components By Polarity-Based Fractionation

Agblevor, F., Utah State University
Ma, B.,

Originally presented on: 11/4/2013 13:45:00 - 14:10:00

Biomass pyrolysis oils can be potential transportation fuels if they could be successfully upgraded and refined. Conventional pyrolysis oils or bio-oils are usually acidic, have high viscosity, low energy density, and are very reactive. Catalytic biomass pyrolysis oils are much more stable but also have acidic properties. In this paper we investigated the fractions of the biooil that are responsible for the aging of the oils through polarity-based fractionation. In this approach column chromatography method was used to fractionate the biooil employing hexane, toluene, chloroform and methanol as eluents. The viscosity of the fractions increased in this order; hexane<toluene<chloroform<methanol. The methanol fraction in most cases was a solid at room temperature and the least viscous oils were the toluene and hexane fractions. Some portions of the chloroform and methanol fractions were also soluble in water. When the water soluble fractions were removed, although the residues were soluble in DMSO and chloroform, they did not produce any signals when subjected to liquid state ^{13}C-nmr analysis. MALDI-TOF analyses revealed that these materials were mostly lignin and carbohydrate oligomers that were soluble in methanol. The increase in viscosity of the biooils was attributed to association of the oligomers and crystallization of levoglucosan in the oil fractions. However, when about 10 wt% methanol was added to the oils the viscosity aging process was reduced because the methanol dissolved the oligomers and prevented the association reaction.

(147d) Distributed On-Farm Bioenergy, Biofuels and Biochemicals Development and Production Via Integrated Catalytic Thermolysis: The FarmBio3 Consortium

Boateng, A. A., USDA Agricultural Research Service

Originally presented on: 11/4/2013 14:10:00 - 14:35:00

Much has been said of farmer-participated distributed processing as the key to economically producing advanced biofuels from cellulosic feedstocks. However, there is much yet to be done to realize the overarching goal of distributed on-farm production of biofuels and chemicals to contribute to the 2022 Energy Independence and Security Act (EISA) goals on advanced biofuels production. Counting on the potential for pyrolysis to provide small footprints that are adaptable to the farm setting, the USDA's Agricultural Research Service's (ARS) pyrolysis program at the Eastern Regional Research Center (ERRC) in Philadelphia has developed a compendium of catalytic and non-catalytic pyrolysis conversion technologies. These endeavors show potential for on-farm pyrolysis to be a truly carbon negative pathway to producing market entry fuels, such as heating oil, gasoline, diesel/jet fuel, and chemical intermediates, but critical barriers still exist. Although village or farm scale pyrolysis production has been modeled, distributed pyrolysis production has not been successfully demonstrated to determine the true economics of an actual integrated system; collective farmer participation is equally unproven and socio-economic and technological barriers still remain. FarmBio^3 is a government-university-industry partnership led by ARS and funded by USDA-NIFA under the Biomass Resource and Development Initiative (BRDI) program. The overarching objective of FarmBio^3 is to leverage existing synergies among partners to address some of these barriers and optimize pyrolysis pathways to commodity fuels and chemicals at various scales following a specific logic model. This presentation will introduce the Consortium members and provide an overview of the work to date. Accomplishments include near complete construction of a two metric ton per day patent-pending dual fluidized bed, combustion-reduction integrated pyrolysis unit (CRIPS), and the development of novel robust catalysts with appropriate chemical selectivity designed for on-the-farm operation.

(147e) Producing Biochars With Consistent Properties From Forestry Residues: Using Spatial Variations of Reactor Temperature History to Choose Pyrolysis Conditions

Gooding, A. A.,
Brewer, C. E., New Mexico State University
Zygourakis, K., Rice University
Originally presented on: 11/4/2013 14:35:00 - 15:00:00

One desirable quality of reactors used for the thermochemical conversion of biomass is that they be "feedstock-blind": capable of producing consistent products even when the feedstocks or the feedstock properties vary. To achieve a satisfactory level of reproducibility, we must be able to accurately control the operation of a reactor so that each biomass load will be processed with the temperature program required to produce biochar with the desired properties.

Biomass pyrolysis involves a complicated series of sequential and parallel reactions that include dehydration, depolymerization and decomposition of biomass components (hemicellulose, cellulose, lignin), devolatilization, gas cracking, condensation, etc. Therefore, the distribution of the pyrolysis products, as well as the chemical and physical properties of the produced biochar, will depend on the complete temperature history of the biomass particles.

For slow pyrolysis of biomass, differences in feedstock properties (moisture, composition, heat capacity, heat of reaction), particle sizes, or flow rates of inert gases change the heat transfer rates and temperature distributions within the reactor. As a result, feedstock and operating differences will have significant effects on the temperature histories of biomass particles.

Past research has consistently shown that biochar properties (like H/C or O/C ratios) vary with highest pyrolysis temperature and reaction time at that temperature. In a recent study[1], we have shown that biomass heating rates and particle sizes also affect the chemical and physical properties of biochars. We need to understand the spatial temperature variations within a reactor in order to be able to make biochars with reproducible and tunable properties.

In this study, wood and residues (branches, leaves, bark) from a managed slash pine and eucalyptus operation were pyrolyzed in a fixed-bed reactor under nitrogen over a range of temperatures and particle sizes. The lab scale (1 L) reactor was custom-built to enable consistent placement of multiple thermocouples to monitor spatial temperature distributions during slow pyrolysis. Overall yields, H/C and O/C ratios, and pore structure characteristics of the produced biochars were evaluated with respect to the internal reactor temperature variability and temperature histories. Finally, we present a theoretical analysis demonstrating how reaction engineering principles and experimental data can be used to develop temperature programs that produce biochars with desired properties from a variety of feedstocks.

[1] Hao Sun, W.C. Hockaday, C.A. Masiello and K. Zygourakis, *Industrial and Engineering Chemistry Research*, **51** (9), 3587–3597 (2012).

(147f) Structure-Activity Relationships in Biomass Chars Formed in An Entrained-Flow Reactor

Newalkar, G., Georgia Institute of Technology
Iisa, K., National Renewable Energy Laboratory
Sievers, C., Georgia Institute of Technology
Agrawal, P. K., Georgia Institute of Technology
Originally presented on: 11/4/2013 15:00:00 - 15:25:00

Pine and switchgrass (sieved to 180-250 μm) were pyrolyzed in an entrained-flow reactor at high temperatures (600-1000 °C) and high pressures (1-20 bar). Heating rates as high as 10^3-10^4 K/s were achieved with solids residence time from 3-40 s. Morphological and chemical changes in these chars are expected to simulate those in industrial reactors. Thus, the gasification reactivity of these chars can be used to model commercial gasifier operations.

Morphological characterization of chars was performed using SEM, N_2 and CO_2 adsorption. CO_2 adsorption was found to measure accurately the total micropore area, which is critical in gasification reactivity measurements. Ultimate and ICP-MS analyses were used to determine the elemental composition. Gasification reactivity of char was measured using TGA at various partial pressures of CO_2, under conditions free of any internal and external mass transport limitations.

Initial gasification reactivity of chars was found to decrease with an increase in pyrolysis temperature and pressure. Although a correlation between surface area as well as ash content with reactivity is observed, it is inadequate to explain the reactivity changes. This is because the drop in reactivity with increase in temperature was much more pronounced than the decrease in surface area or ash content measured over the same temperature range. The intrinsic reactivity may also be impacted by the type of carbon (carbidic vs graphitic) in the char matrix and the active catalytic sites in the char. The type of carbon structure and active sites evolve over the entire conversion range making the reactivity versus conversion profile wavy in appearance. Attempts have been made to explain the reactivity profiles using Avicel char.

Chars were ground to various particle sizes and their reactivity was measured using TGA. The initial reactivity of chars of different particle sizes correlates well with its ash content. This reveals that ash is heterogeneously distributed in the carbon matrix of char. It also reveals that the conventional procedure to measure the pore-diffusion effects by crushing chars can provide misleading results because of the changes in intrinsic reactivity caused by heterogeneous distribution of inorganic content within the different sized char particles. It can be thus seen, that the intrinsic char gasification reactivity is a complex function of the type of carbon, the ash content, ash composition and ash distribution in carbon matrix over the course of the reaction.

(181a) Evaluation of Decay Resistance and Fixation Ability of Catechu and Boric Acid Based Preservative for D. Strictus Bamboo Species

Kaur, P. J., Indian Institute of Technology, Delhi
Pant, K. K., Indian Institute of Technology, Delhi
Satya, S.,
Naik, S., Indian Institute of Technology Delhi
Originally presented on: 11/4/2013 15:15:00 - 15:40:00

In view of depleting wood resources and high cost of metal processing, bamboo has emerged as an excellent substitute. Bamboo species has perfect amalgamation of high tensile strength, low weight to height ratio, hollow tube form structure and relatively high shape factors. This has resulted in high bending strength and torsion, making it most desirable material for the structural design. Also, the high growth rate, and earth quake resistant properties make bamboo a prominent material for advanced engineering and civil construction. However, it is vulnerable to extracellular fungal attack, which restricts its applications. This has circumvented the need for preservation and treatment of bamboo culm. At present, conventional preservatives include solution of Copper Chrome Arsenic (CCA), Copper Chrome Boron (CCB), PentaCholoroPhenol (PCP) etc. Use of Chromium and Arsenic in CCA and CCB resulted in the restriction on use of such preservatives. It has been reported that CCA in preserved wood can rub off onto hands when wood structures are touched and leach from the wood into surrounding soil after exposure to rain and water, thus posing severe health hazards. Therefore, it is impetus to look for alternative preservatives with minimum leaching characteristics.

Natural components extracted from various plant species have been proven to possess antifungal properties. They are used as antimicrobial agents in various traditional medicines. Tannic acid has strong ability to bond with egg proteins, boric acid and form chelates with metal ions. Earlier studies suggest *Pinus radiate*, *Mimosa* plant, *Poria placenta* as a source of tannin with antifungal properties for various wood species. The extract from the heartwood of *Acacia catechu* has been used in India as a preservative for fabrics and fishing nets. This extract is termed “catechu" and has been found to contain a large proportion of low molecular weight flavonoids, catechin in particular.

The bamboo samples were dipped in above mixtures for different duration of time (24, 48 and72 hours). The fixation ability of above solutions, was performed by leaching for 5 days in water using 50 ml water for each specimen, changed each day with a fresh 50 ml volume of water. Folin-Ciocalteau method was used to quantify the tannic acid content in different extracts using Perkin Elmer, Lambda 25, UV-VIS spectrophotometer. Results of leach resistance indicate that leaching efficiency of individual tannic acid, boric acid solution was 44.39%+0.24 and 49.42%+0.12, respectively. Preservative solution based on the cross-linking of condensed tannins complexes with boric acid (5%), Gelatin (1%) and copper oxide (0.1%)

was found to markedly slow down the leaching of boron (leaching:3.2%+0.19).

The bio-efficacy of all treatments was evaluated against *Polyporus versicolor* white rot fungi, using Soil Block Analysis method, as per guidelines of ASTM D 1413. All the treatments have shown decay resistance effectiveness better than control. They significantly enhanced bamboo durability before and after leaching. The catechu extracts treated unleached samples have shown some antifungal properties, with weight loss to be 26.54% +0.23. However, the solution was found to have poor fixation ability inside the bamboo cells. The leached sample with weight loss: 49.12%+0.15 were found to be comparable to control (55.2%). This might be because of water leaching of all the active ingredients from treated products. This will confine the applicability of treated product to limited applications. Addition of boric acid and gelatin has resulted in positive increase in decay resistance. The bamboo sample treated with solution containing boric acid (5%), Gelatin (1%) and copper oxide (0.1%), weight loss of leached and unleached samples was found to be 9.25+0.23 and 8.25%+ 0.19 respectively. The ability of tannic acid to bond with gelatin and boric acid, might have helped preservative to be bonded to the cells of bamboo culm, resulting in drastic reduction in leaching.

The above research findings are very encouraging. Results show a strong possibility of alternative treatment technique to synthetic chemical preservatives. Overall, it represents an advanced strategy capable of imitigating both environmental and health concerns associated with the prevailing trends in bamboo preservatives.

(181b) Removal and Recovery Of Dye From Paper Machine Efffluents By Continuous Column Method

Ray, A. K., Indian institute of technology, Roorkee
Utgata, T., Century Pulp & Paper
Mohanty, P. K., Indian institute of technology, Roorkee
Originally presented on: 11/4/2013 15:40:00 - 16:05:00

REMOVAL AND RECOVERY OF DYE FROM PAPER MACHINE EFFFLUENTS BY CONTINUOUS COLUMN METHOD

Mohanty* P., A.K. Ray and T. Utgata**

Department of Paper Technology

Indian Institute of Technology

Saharanpur, India

* Ballarpur Industries Ltd., Orissa

** Century Pulp & Paper, Lalkua, Uattar Preadesh.

ABSTRACT

Various kinds of paper such as colored poster paper,adverstising etc. use dyes and color pigments. These color components are generally non-biodegradable and affect environment severely when goes to the water bodies. Therefore it needs removal of these dyes. Adsorption of dye on a suitable adsorbent is a plausible alternative of abating colour (dye) pollution from paper machine effluents. Experimental studies have been carried out using Methyl violet and Victoria blue dyes as adsorbates, on adsorbents like bagasse pith, wood meal and crushed bleached pulp. The experiments were performed in both batch as well as continuous adsorber. For batch process parametric effects of particle size, agitation, initial dye concentration, time, and temperature on the rate of dye removal have been investigated. Detailed results of batch process were reported earlier. Effects of surface mass transfer, intra-particle diffusion on rate of dye removal were correlated in terms of Schmidt and Sherwood number.

In this present study the result of continuous process are highlighted. Comparative results of batch as well as continuous process are also shown. Overall as well as individual effects of time, temperature and bed height on dimensionless concentration are mainly studied. The continuous adsorption process are performed for a particular bed height in the range between 3 cm and 11 cm. For each bed height an adsorbate- adsorbent combination is selected and adsorption is carried out at three different temperature ranges, namely 303 K,318 K and 333K. For effects of above parameters ,Initial concentration is however kept at 2 gpl and the particle size of all adsorbates at 40 mesh. Confirmatory trial experiment is performed using methyl violet as adsorbate at 1 gpl concentration with wood meal as adsorbent at 303 K-333K.

(181c) Economic Production Of BIO-Ethanol From Various Indigenous Sources In India-A Review

RAY, A. K., IIT Roorkee
Yadav, S., Iit Roorkee
Sharma, P., Indian Institute of Technology Roorkee
Originally presented on: 11/4/2013 16:05:00 - 16:30:00

Ethanol is usually produced in India by fermentation of molasses (generally 5% on cane) with a yield, 250.7 liter per tons of molasses with yeast *Saccharomyces cerevisiae* or by strains, *Zymomonas mobilis*. Uses of cane juice, jaggerry, mohua flower are also used in some installations. An alternative feedstock for ethanol production is ligno-cellulosic biomass;bagasse ,rice straw, cotton straw, groundnut shells, wood and forest residues, waste paper products, paper and pulp mill wastes using thermophilic anaerobic bacteria, *Clostridium thermocellum*. Cellulose enzyme produced by aerobic mesophilic fungus, *Trichoderma reesei, also* releases glucose from plant biomass and converts to ethanol by simultaneous saccharification and fermentation.

Acid hydrolysate of hemicelluloses extracted from bagasse and other agriculture wastes by fermentation with the yeast *Saccharomyces cerevisiae*has also been a potential feedstock.

Starch based natural produce like wheat, rice, millets, potatoes, corn, tapioca can also be a potential and economic source of ethanol production. In this paper potential for profitable production of alcohol by various indigenous sources are reviewed. Economics of alcohol manufacture in relation to various constraints is discussed.

(181d) Strategies For High Vacuum Dewatering On Paper Machines

Ray, A. K., Indian institute of technology, Roorkee
Pandey, R.,

Originally presented on: 11/4/2013 16:30:00 - 16:55:00

The Paper machine vacuum system is an important contrivance for dewatering of wet web as well as increasing percentage solids in web.There are various models available to assess the influence of vacuum on the above parameters. In this present investigation the steady state model is discussed for estimating optimum vacuum load in flat boxes of fourdrinier paper machine as the flat boxes are the main components to dewater the sheet as well as wire/ felts in the paper machine. From the detailed investigation it has been found that for a given furnish and vacuum level, there is an ultimate achievable dryness beyond which the sheet will not go, no matter what the length of exposure is.Further there are two main factors in vacuum dewatering - the ultimate achievable dryness for that vacuum, and the rate at which that dryness is attained. This paper also indicates the effect of temperature and slots design for vacuum dewateringThe Paper machine vacuum system is an important contrivance for dewatering of wet web as well as increasing percentage solids in web.There are various models available to assess the influence of vacuum on the above parameters. In this present investigation the steady state model is discussed for estimating optimum vacuum load in flat boxes of fourdrinier paper machine as the flat boxes are the main components to dewater the sheet as well as wire/ felts in the paper machine. From the detailed investigation it has been found that for a given furnish and vacuum level, there is an ultimate achievable dryness beyond which the sheet will not go, no matter what the length of exposure is.Further there are two main factors in vacuum dewatering - the ultimate achievable dryness for that vacuum, and the rate at which that dryness is attained. This paper also indicates the effect of temperature and slots design for vacuum dewatering

(147g) Formation of Pyrolytic Lignin During Fast Pyrolysis of Biomass

Kim, K. H., Iowa State university
Bai, X., Iowa State University
Dalluge, E.,
Hutchinsone, C.,
Dalluge, D., Iowa State University
Lee, Y. J.,
Brown, R., Iowa State University
Originally presented on: 11/4/2013 15:15:00 - 15:40:00

Bio-oil obtained from fast pyrolysis of lignocellulosic biomass usually contains up to 30% of pyrolytic lignin, which mainly consists of phenolic oligomers. Phenolic oligomer is undesired because it contributes to thermal instability of bio-oil and catalyst deactivation in bio-oil upgrading. Formation mechanisms of phenolic oligomers and how the high molecular weight compounds are recovered in bio-oil are not well understood. Whether it is lignin fragments that are thermally ejected as aerosols or repolymerization products of phenolic monomers is still under debate. In the present study, lignin is fast pyrolyzed in a micropyrolyzer under conditions that limit secondary reactions. The pyrolysis vapor is analyzed using GC/MS and solvent-recovered (SR) liquid from the micropyrolyzer is analyzed using high resolution Fourier Transform Ion Cyclone Resonance MS (FT-ICR-MS) and orbitrap-MS utilized by Atmosphere Pressure Photoionization (APPI) to obtain nearly complete information about pyrolysis products. The origin of the products recovered from the micropyrolyzer is further investigated by pyrolyzing several phenolic monomers and dimers at the same experimental condition.

The results show that the pyrolysis products of lignin under conditions that suppress secondary reactions consist of phenolic monomers and oligomers up to molecular weight of 504. It is also found that primary phenolic monomers of lignin pyrolysis rapidly repolymerize and thermally crack via side chain rearrangement. At temperatures up to 500°C, repolymerization is stronger than thermal cracking. The original structural components of lignin, such as phenylcoumaran, stilbene and biphenol decompose during fast pyrolysis. However, the extent was incomplete and most of their derivatives are still dimers. Therefore, phenolic dimers representing fragments of the original lignin can be primary products of lignin depolymerization during fast pyrolysis. On the other hand, resinol structure is thermally unstable and does not survive fast pyrolysis. These results suggest that primary products of lignin pyrolysis are mainly monomers and dimers. Most fractions of phenolic oligomers found in bio-oil are the repolymerization products during the vapor-phase recovery and bio-oil storage.

(207a) Time-Resolved Fast Pyrolysis Using a Free-Fall Reactor and a Time-of-Flight Mass Spectrometer

Johnston, P., Iowa State University
Gable, P.,
Brown, R., Iowa State University
Originally presented on: 11/4/2013 15:40:00 - 16:05:00

Time resolved fast pyrolysis was investigated using a Time-of-Flight Mass Spectrometer (TOF-MS) integrated with a Free-Fall Reactor. Sampling ports along the axis of the vertically oriented reactor provided spatially resolved measurement of chemical species according to molecular weight. Spatially resolved data was converted to time resolved concentration based on free-fall velocities of the pyrolyzing biomass. In some cases it was possible to assign molecular weights to specific molecular species although for most lignin-derived compounds this was not possible. This presentation focuses on the anhydrosugar levoglucosan, which is easy to identify from the MS data and has a relatively well-known reaction pathway, although little is known about the kinetics of its formation. The time-resolved data for levoglucosan is used to estimate reaction rates for its formation from cellulose. The effects of reactor temperature, sweep gas velocity, and biomass-to-sweep gas ratio are evaluated.

(207b) The Nature of Torrefaction Effect On Producing High-Quality Fuel Intermediate Via Fast Pyrolysis

Meng, J., North Carolina State University
Park, J., North Carolina State University
Moore, A., North Carolina State University
Tilotta, D., North Carolina State University
Adhikari, S., Auburn University
Kelley, S., North Carolina State University
Park, S., North Carolina state university
Originally presented on: 11/4/2013 16:05:00 - 16:30:00

Session T4 08 Thermochemical Conversion I

AIChE 2013

November 3 ~ 8, 2013

San Francisco, CA

The Nature of Torrefaction Effect on Producing High-Quality Fuel Intermediate via Fast Pyrolysis

Jiajia Meng,[1] Junyeong Park,[1] Andrew Moore,[1] David Tilotta,[1] Sushil Adhikari,[2] Stephen Kelley,[1] Sunkyu Park[1]

[1] Department of Forest Biomaterials, North Carolina State University, Raleigh, NC 27695

[2] Biosystems Engineering Department, Auburn University, Auburn, AL 36849

*Presenting author: jmeng@ncsu.edu

As a new rising pretreatment technology for making high quality bio-oil via fast pyrolysis, torrefaction pretreatment has attracted increasing attention last years. This simple thermal pretreatment was expected to reduce oxygen content as well as associated water and acids in the resulting bio-oil, therefore bringing significant benefits (reducing hydrogen consumption and coking formation) to the further oil upgrading process, i.e. hydrotreating. Understanding the effect of torrefaction treatment on bio-oil properties and pyrolysis chemistry is critical to fulfill these expectations. In this study, fluidized-bed fast pyrolysis was performed on torrefied loblolly pine with different degrees of torrefaction severity. With the bio-oils made from this integrated process, the effect was investigated via: (1) comparing the physical and chemical properties, including oxygen content, water content, viscosity, average molecular weight, acidity (TAN number) heating value and chemical composition (determined by GC/MS and solvent fractionation) of bio-oil made from torrefied wood to that made from non-treated wood; (2) semi-quantifying the carbon and hydrogen distribution between the bio-oils made from the pretreated wood and the raw wood with ^{1}H and ^{13}C NMR characterization; (3) accessing the time and thermal stability of bio-oil made from pretreated wood by accelerated aging and natural aging methods. The results show that torrefaction pretreatment can significantly reduce the oxygen and water content of the bio-oil as well as varying the chemical composition of bio-oil with concentrated pyrolytic lignin and levoglucosan. Less polymerization occurred during the aging process for bio-oil made from torrefied wood. All these effects are attributed to the structural and compositional change of torrefied wood induced by the thermal treatment. Possible cellulose pyrolysis mechanism was proposed to explain the increased levoglucosan concentration in the torrefaction-based bio-oil. In addition, the torrefaction effect on fast pyrolysis process was accessed by separated and integrated torrefaction and pyrolysis mass balance, elemental balance, and energy balance. With these understanding, an integrated process, including torrefaction pretreatment, fast pyrolysis, and oil upgrading to produce hydrocarbon fuel, is proposed.

(207c) Upgrading of Biomass Pyrolysis Vapor
Nimlos, M. R., National Renewable Energy Laboratory
Iisa, K., National Renewable Energy Laboratory
Mukarakate, C., National Renewable Energy Laboratory
Robichaud, D., National Renewable Energy Laboratory
Jarvis, M. W., National Renewable Energy Laboratory
Originally presented on: 11/4/2013 16:30:00 - 16:55:00

Fast pyrolysis of biomass shows great potential as a means of efficiently producing carbonaceous liquids that can be used as feedstocks for the production of transportation fuels and chemicals. However, the resulting oils require extensive hydrotreatment to improve their properties and upgrading the vapors before they are condensed should be considered. This presentation will discuss ongoing work at the National Renewable Energy Laboratory (NREL) to investigate pyrolysis vapor upgrading. In particular, we are interested in how the catalysts and reactor conditions for the upgrading step affect the composition and properties of the resulting oil. We have used laboratory reactors to study catalytic transformation of biomass pyrolysis vapors and have characterized the products in the gas phase using Molecular Beam Mass Spectrometry (MBMS) and GC/MS. We have also condensed oils from packed bed, entrained flow and fluidized bed reactors. The composition and physical properties of the oils produced from these reactors and their comparison to the gas phase measurements will be presented. These results will be discussed with regard to the viability of the entire conversion process, including potential refinery introduction.

(207d) Catalytic Pyrolysis of Torrefied Biomass Using Different Acidic Zeolites

Adhikari, S., Auburn University
Srinivasan, V., Auburn University
Park, S., North Carolina state university
Originally presented on: 11/4/2013 16:55:00 - 17:20:00

The liquid fuel derived from pyrolysis of biomass exhibits certain characteristics that limit its usage as an efficient transportation fuel. Several upgrading methods to produce aromatics and to selectively eliminate oxygenates derived from the pyrolysis process are available, among which catalytic pyrolysis is gaining a lot of research attraction lately. Fast pyrolysis of biomass using zeolite catalyst has shown to be effective in improving the aromatic production. This study focuses on the aromatic production through catalytic pyrolysis of biomass and its two model compounds--cellulose and lignin. Pre-treatment (torrefaction) was performed to observe its effect on pyrolysis products. The influence of SiO_2/Al_2O_3 ratio of zeolite catalyst on selectively eliminating oxygenates during the pyrolysis of raw and torrefied model compounds were investigated.

(207e) Hydrogen From Acetic Acid As the Model Compound of Bio-Oil Over Ni/Ce-Zr-O Catalyst

Yan, C., Guangzhou Institute of Energy Conversion, Chinese Academy of Sciences

Originally presented on: 11/4/2013 17:20:00 - 17:45:00

The main method available for the hydrogen production from biomass is thermo chemical conversion of biomass to hydrogen, including steam gasification[4, 5] high-pressure supercritical conversion and steam-reforming of bio-oils. Bio-oil is a liquid by fast pyrolysis of biomass, composed by a complex mixture of oxygenated organic molecules and water. Due to its ease of storage and transport, as well as higher energetic density, hydrogen production by steam reforming of bio-oil is recognized as one of the most promising and economical ways to producing hydrogen.

Ni/Ce-Zr-O catalysts were prepared with the different methods of co-precipitation, impregnation and impregnation-co precipitation, and the content of nickel and cerium are 12wt% and 7.5wt%, respectively. Structures and reduction properties as well as surface morphologies of the catalysts were characterized by means of BETXRDAH_2-TPR and SEM. Results showed that crystal particle size the Ni/Ce-Zr-O catalyst prepared by co- precipitation method was uniformer and pores of that were richer, which made the activity of catalyst for hydrogen production from acetic acid be better than the Ni/Ce-Zr-O catalysts prepared by impregnation method and by impregnation-co precipitation method obviously. When the reaction temperature is 700°C, steam to carbon is 3, and the liquid hourly space velocity is 4.7 h^{-1}, the hydrogen yield of Ni/ Ce-Zr-O catalyst prepared by co- precipitation method can reach 83.4% of the hydrogen yield from steam reforming of acetic acid.

(207f) Separations of Organic Acids From Pyrolysis Bio-Oil

Lin, Y. J., Argonne National Laboratory
Sather, N.,
Libera, J., Argonne National Laboratory
Snyder, S., Argonne National Laboratory
Originally presented on: 11/4/2013 17:45:00 - 18:10:00

Organic acids, such as acetic acid and formic acid, are the major components in bio-oil after water and hydroxyacetaldehyde. These organic acids and inorganic salts are formed as byproducts during pyrolysis. Stability and acidity of bio-oil have been major technology hurdles that cause problems in storage and transportation of pyrolysis bio-oil and the causes are attributable to the combination of water, mineral salts and organic acids contained with other organic constituents of bio-oil. The carboxylic acid group also influences the quality of intermediates for bio-oil upgrading to hydrocarbon fuels. Therefore, removing and converting the acid from bio-oil into other stable intermediates for upgrading to hydrocarbon fuel will provide solution to address multiple technical hurdles. In this presentation, we will discuss the effects of organic acid removal from pyrolysis bio-oil on the bio-oil stability and acidity. Water extraction was used to stabilize the oil phase of bio-oil by removing water soluble compounds, including the acids, into the aqueous phase. Near 70 wt.% of the biomass was in the aqueous phase. We will discuss the results of acids recovery from the aqueous phase of bio-oil using a proprietary resin wafer electrodeionization (RW-EDI) technology. Different approaches to convert the recovered organic acids in aqueous phase into stable intermediates for upgrading to hydrocarbon fuel will be presented.

(221a) Preparation of Acetylated Wood With Solvent-Free Gas Phase Method

Monday, November 4, 2013
Grand Ballroom B (Hilton)

Xiaoli GU, Xu MA and Lixian LI, College of Chemical Engineering, Nanjing Forestry University, Nanjing, China

(221a) Preparation of acetylated wood with solvent-free gas phase method

XIAOLI GU*[1,2], XU MA[1,2], LIXIAN LI[1],
(1 Nanjing Forestry University, College of Chemical Engineering, Nanjing CHINA
2 State Key Laboratory of Pulp and Paper Engineering , South China University of Technology , Guangzhou CHINA)

KEYWORDS: poplar; gas phase method; wood acetylation; acetic anhydride;

For wood acetylation as chemical modification, the agent of acetic acid reacting with molecules of wood on hydroxyl, hydrophobic acetyl replaced the hydrophilic hydroxyl. Acetylated wood have advantages such as uniform density, smooth surface, dimensional stability, strong corrosion resistance, low heat ductility and decreasing the toxicity, etc.

Wood powder and sawdust acetylation were reported in the 20th century (Guan 2007, Jianzhang et al. 2002), but the research of real wood acetylation reports began in 1946. Then the United States, Japan and other countries of the world enterprises, universities and research institute successively started research the wood acetylation technology and commercial application.

Current research found low acetylation agent efficiency, e.g., about 85% of anhydride not participating in the reaction (Kazuya et al. 2004, Yoriko et al. 2005, Eiichi et al. 2009, Mohamed et al. 2008). Based on the actual problem, we successfully took the advantages of gas phase acetylation with solvent-free process.

Poplar wood (size 20 mm x 20 mm x 10 mm), produced in Jiangsu, with 105 °æ drying for 24 h was applied as substrate in the experiments. A DHG-9146 type electro-thermal constant temperature drying oven and a Nicolet FT-IR infrared spectrometer were used as characterization instruments.Pieces of wood samples with 105 °æ drying for 24 h before the experiment. Acetylation processing temperature was set for 90-175°æ and the processing time was set for 0-10 h.

$$\text{Weight Percent Gain} = \frac{m_{(t)} - m_0}{m_0}$$

$m_{(t)}$ was the experimental mass at each monitoring time, and m_0 was the initial dry mass.

Figure 1A showed the effects of the reaction temperature and reaction time on the WPG of treated samples. The WPG increased with the increment of reaction temperature and reaction time. The acetylated wood measured 6.69% WPG at 1h and 14.92% WPG at 2h when the reaction temperature was 140°æ. Then the acetylated wood showed a small weight gain with the increase of temperature and the WPG was 18.79% at 4h. At 140°æ, the WPGs of the acetylated wood increased quickly with the reaction period from 1h to 2h.

At 150°æ,the WPG was only 0.29% more than that at 140°æ when the reaction time for 2h. At 150°æ, the acetylated wood was measured 19.73% WPG at 4h,which was 0.94% more than that at 140°æ in the same reaction time. Therefore, the improvement of the reaction temperature, the effect of promoting initial stage of acetylation reaction was obvious.

At 160°æ the WPG of the acetylated wood increased quickly than that at 150°æ, e.g., the WPG was 18.87% at 2h. The acetylated wood showed a small weight gain after reacting for 3h. The WPG of two periods from 3h to 4h were 0.05%, 0.11% respectively. It was shown that the reaction time was not the prominent factor for the increase of acetylated rate.

The WPG of the acetylated wood increased obviously in the reaction temperature reached 170°æ above. At 170°æ, as the reaction time increased from 2h to 4h, the WPG of acetylated wood increased from 21.82% to 24.07%. At 175°æ, the WPG of acetylated wood increased from 21.92% to 26.09% at the same conditons. Thus it can be seen, after reaction temperature 170°æ above, increasing the temperature leaded to the WPG increase apparently, which was related to a tendency to the carbonization of the acetylated sample as the temperature increased.

In order to show the superiority of gas phase method, we carried on liquid acetylation experiments for comparison. The experimental results (figure 1B) showed that the WPG reached the maximum value at 8h in liquid phase and the acetylated wood showed a small decrease trend when continual increasing the temperature.

At 110°æ, the WPG was 8.3% at 2h in liquid phase, which was 0.3% less than that at 170°æ for 0.5h in gas phase. At 120°æ, the WPG was 17.91% at 8h in liquid phase, which had a certain increase in small degree as compared with that at 110°æ. At 120°æ, as the reaction time increased from 4h to 8h, the WPG of acetylated wood increased from 12.36% to 17.91%.

When the reaction temperature reached 130°æ,the increase of WPG was obvious as the reaction time increased. The WPG was 22.63% at 8h. Under the conditions of 130°æ, the excretion of massive acetic acid as the by-products of acetylation promoted the increase of WPG in some degree. In addition, we also found that the most obvious time of WPG increased was from 4h to 8h during liquid phase acetylation.

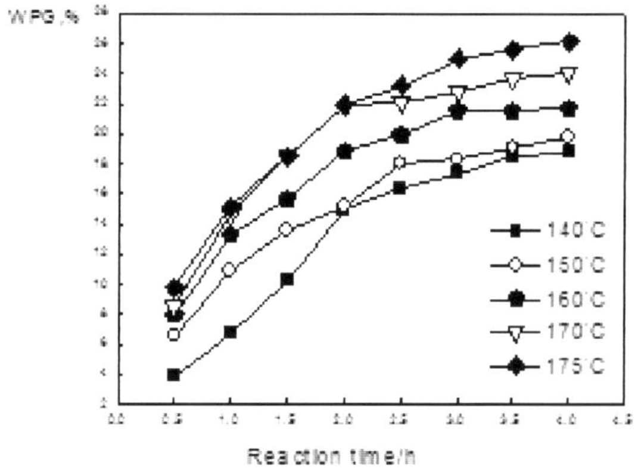

Fig 1A weight percent gain of gas phase acetylation

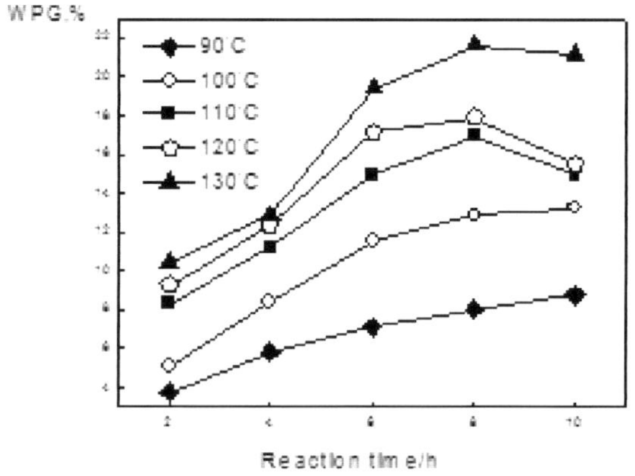

Fig1B weight percent gain of liquid phase acetylation

The absorption peaks of related groups in acetylation are: -OH absorption peak at 3400 cm^{-1}, saturated esters carbonyl absorption peak at 1742 cm^{-1}, -CH of acetate, absorption peak at 1384 cm^{-1}, C-O of acetyl absorption peak at 1248 cm^{-1}. The absorption peaks degree of related groups of acetylated wood were strengthened obviously as shown in figure 2A. And integration of FTIR data of several related absorption peaks resulted was also shown in figure 2B:

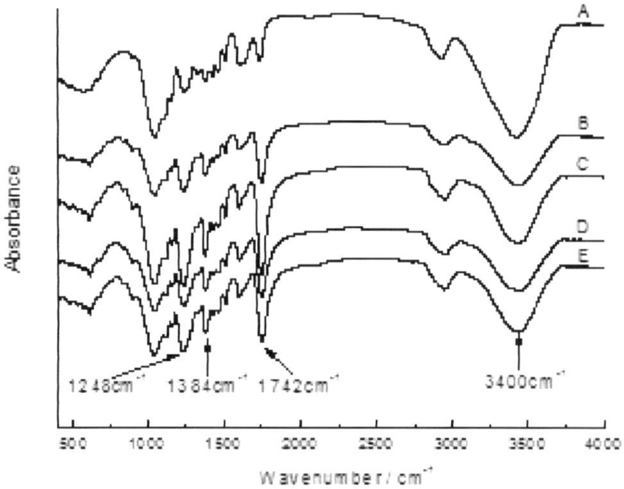

Fig 2A Comprehensive comparison of different treatment acetylation samples

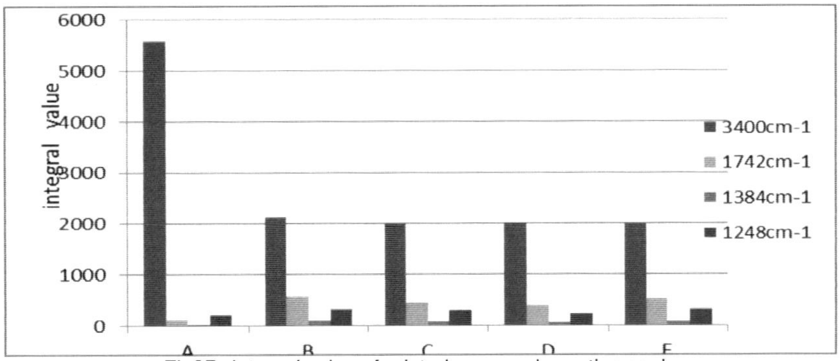

Fig2B Integral value of related groups absorption peaks

Note: A) the substrate sample; B) the acetylated wood with liquid phase acetylation at 90°æ for 8h; C) the acetylated wood with liquid phase acetylation at 130°æ for 8h; D) the acetylated wood with gas phase acetylation at 140°æ for 4h; E) the acetylated wood with gas phase acetylation at 175°æ for 4h.

The hydroxyl absorption peak at 3400 cm^{-1} had largely reduced after acetylation as shown in Fig2B, the amount of reduced hydroxyl was above 60%. Because of some water content in the wood, the amount of reduced hydroxyl can't reach zero. As compared with the substrate sample, we can see the acetylation significantly occur from several of acetyl characteristic peaks , especially the absorption peak of saturated esters carbonyl at 1742 cm^{-1} increased obviously, average amount up to 300 times or so, e.g., the maximum group can reach 373.76 times. Meanwhile, the average amount of the absorption peak of -CH in the acetate at 1384 cm^{-1} increased up to 158 times and the average amount of the absorption peaks of acetyl C-O at 1248 cm^{-1} up to about 50 times.

In conclusion, With the pure acetic anhydride as acetylation agent for the poplar wood acetylation process, the influence of the processing temperature (140-175°æ) and the processing time (2-4h) on the acetylation yield was investigated. Experimental results showed that as the temperature and the reaction time increase, the acetylation yield increase gradually. The acetylation yield was above 26% with the temperature of 175°æ for 4 hours. The acetylation rate is 2.27%/h in liquid phase acetylation and the acetylation ratio of gas phase is 7.21%/h, which was 3.18 times of the former method. Therefore, the gas phase method is a high yield of acetylation method with the advantage of pure acetic anhydride as reagent, not adding other solvents and short reaction period.

REFERENCES

1. Guan, N., 2007: Wood acetylation and commercial application. International Wood Industry 37(4):31-32.
2. Jianzhang L.,Wenrui Z.,Yongjun Q., 2002: Foreign wood acylation research and application progress. International Wood Industry 32(4):9-11.
3. Kazuya M., Yuko I., 2004: Analysis of the factors influencing the acetylation rate of wood. Journal of Wood Science 50:519¨C523
4. Yoriko I., Takafumi I., 2005: Vapor phase reaction of wood with maleic anhydride I: dimensional stability and durability of treated wood. Journal of Wood Science 51:595¨C600
5. Eiichi O., Kazuya M., 2009: Potassium acetate-catalyzed acetylation of wood at low temperatures I: simplified method using a mixed reagent. Journal of Wood Science 55:18¨C22
6. Mohamed J., Gilles S., 2008: A new process for the esterification of wood by reaction with vinyl esters. Carbohydrate Polymers 72:657¨C663

(221b) Correlation of Expression Patterns in Wood-Degrading Brown Rot Fungi to Wood Modifications: Using Space As a Gauge for Time

Presley, G.,

Brown rot fungi consolidate enzymatic bioprocessing with a non-enzymatic pretreatement that could be chemically mimicked in an engineered process. Wood degraded by brown rot fungi is characterized by a rapid depolymerization of cellulose while lignin is primarily demethylated and left in place. Brown rot fungi accomplish this by producing a diffuse, hydroxyl radical-producing Fenton system at a distance from hyphae and discrete from secreted cellulases. This co-localization of free radicals and hydrolytic enzymes is intuitively not feasible, yet it is a hallmark of the brown rot mechanism, as theorized. Previous work spatially characterizing wood wafers decayed by *Postia placenta* indicate that depolymerization of wood polysaccharides preceded the secretion of endoglucanases as well as lignin modifications, indicating a role for an unknown cellulolytic mechanism during early colonization. My work is focused on spatially mapping gene expression along the advancing hyphal front. Unlike previous analyses comparing expression as a function of substrate, my goal is to create a temporal series by sampling along a gradient of directional growth. This system will be further investigated using RT-qPCR and RNA-seq to identify genes important in incipient wood decay stages. This will allow the study of the colonization of intact wood at a much finer timescale than previous expression studies in brown rot fungi, and help identify the mechanisms brown rot fungi use to extract glucose from wood without first removing lignin.

(221c) Development and Validation of a Hemicellulose Depolymerization Kinetic Model for Aspen During Dilute Acid Hydrolysis

Fan, J., Michigan Technological University
Shonnard, D. R., Michigan Technological University

Research into ethanol production from lignocellulosic biomass has grown significantly over the last few decades, for it offers a potential solution to replace conventional fossil fuels while not competing with food production. Hemicellulose is a major component of lignocellulosic biomass, accounting for 25-35% of total dry mass. Thus, the efficient conversion of hemicellulose to fermentable sugars is vital to ethanol yield and optimizes the economic performance of the production process. A kinetic mechanism for hemicellulose hydrolysis is highly desired as a tool to understand and improve lignocellulosic biorefining and therefore, it continues to be modeled by researchers around the world. The most common mechanism is a two-step pseudo first order irreversible reaction where xylan in hemicellulose is hydrolyzed directly to xylose, which is dehydrated subsequently to furfural and eventually tars. However, oligomers are found to be a significant fraction of the product for dilute acid pretreatment, especially at short times, and oligomers are not taken up by fermenting microorganisms unless the oligomers are hydrolyzed further. In this study, a depolymerization model proposed by Lloyd and Wyman (Lloyd and Wyman 2003) is applied to describe the dilute acid hydrolysis of hemicellulose. Aspen samples were hydrolyzed by 0.5 wt% dilute acid at 150, 160 and 175°C. A HPLC coupled with the Hi-Plex Na column was used to measure the concentrations of xylose, xylooligomers, furfural and hydroxymethyl furfural (HMF) along the reaction time path. The depolymerization model successfully predicts the xylose profiles in dilute acid hydrolysis. The reaction is faster in higher reaction temperature as indicated by the increasing hydrolysis rate k_h. Simultaneously, xylose is subjected to more severe degradation process, which is indicated by the increasing degradation rate k_d, which is confirmed by the increasing concentrations of the degradation products furfural and HMF. The kinetic model and experimental data agree on many aspects of the reaction trends; i. increasing temperatures lead to decreased time required for peak concentrations on all oligomer species, and ii. the time to reach peak concentrations of each oligomer decreased with degree of polymerization for all temperatures. However, the experimental data of xylooligomers were much lower than the model prediction, which may result from the model itself or the experiment data. New equipment and experiment procedures need to be implemented to obtain more accurate measurements of xylooligomers. The depolymerization model also needs to be modified to include direct degradation of xylobiose and xylotriose. In addition, the assumption that all bonds react at equal rates can be modified to include differences in end bonds, and also that hydrolysis rate depend on the chain length of xylooligomers.

References

Lloyd T, Wyman CE. Application of a depolymerization model for predicting thermochemical hydrolysis of hemicellulose. Applied Biochemistry and Biotechnology 2003;105:53-67

(221d) Dissolution Processing of Cellulose

Ghasemi, M., University at Buffalo - The State University of New York (SUNY)
Tsianou, M., University at Buffalo - The State University of New York (SUNY)
Alexandridis, P., University at Buffalo - The State University of New York (SUNY)

Nanoscale and mesoscale order in polymers, manifested by chain crystallinity and organization in spherulites or fibers, significantly affects the dissolution of nanostructured polymers, and constrains their subsequent physical or chemical processing. Cellulosic biomass provides a nice example of a composite material that is recalcitrant to solvents. The efficient utilization of cellulose as a starting material for the synthesis of high value-added functional polymers and chemicals and also for biofuel production provides the motivation for the current project. We consider here the transport phenomena governing the dissolution of solid cellulose, e.g., solvent penetration, transformation from crystalline to amorphous domains, specimen swelling, and polymer chain untangling, and analyze the effect of various parameters on the kinetics of dissolution.

(221e) Kinetic on Microwave-assisted Delignification of Wheat Straw by Caustic Soda Solution

Deng, Y., School of Material Science and Chemical Engineering, Tianjin University of Science and Technology
Zhang, Z. M.,
Chen, W.,

Under the condition of atmospheric pressure and different microwave radiation power, cooking experiment were conducted. Pulp yield and lignin content was increasing along with the microwave power growing, pH value under 400W was the highest. Then the mechanisms and kinetics of NaOH pulping of wheat straw were studied. The process condition was set as follows: microwave radiation power was 400 W, NaOH solution content was 1%, solid-liquid ratio was1:10, and the microwave heating temperature was from 50 ℃to 95 ℃, the time was from 10 mins to 45 mins. The result shows that the rate of delignification reaction was 1.3 order with respect to residual lignin and 0.21 order with respect to [OH-]. The activation energies of the delignification was 25.4 kJ/mol.

(234a) Fluorescence Properties of CMC/Eu Nanocomposites in Different Phs

Li, W. H., South China University of Technology
Ye, J., South China University of Technology
Xiong, J., Sichuan University
Originally presented on: 11/5/2013 8:30:00 - 8:55:00

CMC/Eu nanocomposites were synthesized successfully by reacting with Eu^{3+} and CMC(70°¦, 15min) in waterbath. The FTIR results indicated the reaction between carboxyl groups in CMC and Eu^{3+}, and SEM results showed that the particles of CMC/Eu nanocomposites were sphere and particle sizes were around 100 nm. In addition, effect of pHs on the fluorescence properties of CMC/Eu nanocomposites was complicated (see Figure). The emission spectra had red shifts with increasing pH. The fluorescence intensities of the materials in lower pHs were significantly stronger than that of materials in higher pH. It might result from the degradation of ligand CMC in strong alkaline condition.

Acknowledgement: Supported by the National 973 project under Grant No. 2010CB732201, the National Natural Foundation of China under Grant No 31270617 and the Opening Project of State Key Laboratory of Polymer Materials Engineering (Sichuan University) under Grant No. KF201301.

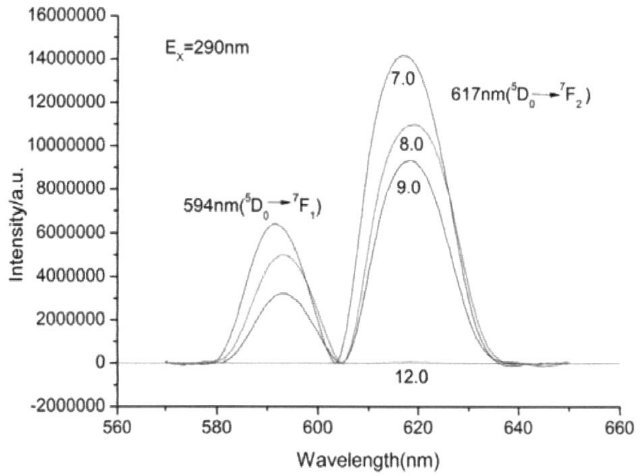

Figure. Emission spectra of CMC/Eu nanocomposites synthesized in different pHs.

(234b) High Performance Green Barrier Films From Thermal Treatment of Cellulose Nanofibrils

Sharma, S., Georgia Institute of Technology
Hu, Z., Georgia Institute of Technology
Zhang, X., Georgia Institute of Technology
Nair, S. S., Forest Products Lab
Ragauskas, A. J., Institute of Paper Science and Technology, Georgia Institute of Technology
Zhu, J., USDA Forest Products Laboratory
Deng, Y., Georgia Institute of Technology
Originally presented on: 11/5/2013 8:55:00 - 9:20:00

Hornification due to thermal exposure has been shown to be detrimental to the properties of cellulose fibers and the formation of paper. This is due to closure of pores, reduction in hydrophilic nature of fibers due to irreversible hydrogen bonding of the –OH groups of the fibers and increased crystallinity. However, these effects of thermal exposure could be very beneficial in the enhancement of barrier properties of membranes made from Nanocellulosic fibers (NCF) fibers for oxygen and water vapor. In this study we demonstrate that controlled thermal exposure (heat treatment/annealing) after forming membranes from NCF fibers can indeed enhance the barrier properties of these membranes. We demonstrated 3 hours of thermal exposure (at 175°C) to films of about 75μm thickness results in the improvement in oxygen barrier by almost 25 fold while water vapor permeability could be reduced to half, while being accompanied by all the consistent hornification changes. The films were found to be decreasing in porosity with increasing heat treatment temperature. The oxygen and water vapor permeability reduced from 0.17 to 0.007 (ml.μm/m2.day.kPa) and 3.12 to 1.544 (kmmol.μm/m2.day.kPa) respectively. Thermal exposure to the films was also accompanied by expected hornification effects, closure of porous structure, increase in size of crystallites by more than 45% (0.067 – 0.099nm) increase in hydrophobicity increase in contact angle from 64° - 96° and decrease in water retention value by more than 55%. This study shows that the barrier and mechanical properties of films made from NCFs can be fine-tuned by controlled thermal exposure.

(234c) Bio-Composite of PLA-b-Pdmaema Copolymer and Carboxymethyl Cellulose (CMC) With Tunable Properties

Tong, Z., University of Florida
Chen, N.,
Originally presented on: 11/5/2013 9:20:00 - 9:45:00

Abstract: CMC-based nanocomposites have been mainly limited by its hydrophilicity, leading to incompatible with hydrophobic polymers, such as PLA (poly lactic acid). In this project, the quaternized and hydrophilic PDMAEMA (poly 2-dimethyl-amino ethyl methacrylate) is proposed to be a compatibilizer between hydrophilic CMC and hydrophobic PLA. Meanwhile, the introduction of soft domain (PDMAEMA micelle structure) can play a role as lubricant between stiff micro-fibrils and PLA for energy dissipation. The properties of obtained composites can be tuned by changing CMC mass ratio, copolymer mass ratio and adjusting molecular weight of PLA or PDMAEMA. Various molecular weight of PLA-b-PDMAEMA copolymers were prepared through ring opening polymerization and following ATRP (atom transfer radical polymerization) reaction. The successful synthesis of PLA-b-PDMAEMA with different molecular weight was confirmed by ^{1}H-NMR (nuclear magnetic resonance) and size exclusion chromatograph (SEC). Then, quaternization was conducted on PDMAEMA to obtain the cationic PLA-b-PDMAEMA copolymer, which can be dispersed into water forming micelle structure. Its hydrodynamic radius was determined by dynamic light scattering (DLS). After that, different mass ratio of cationic PLA-b-PDMAEMA and anionic CMC were mixed together through ionic interaction to form biomimetic nanocomposite. The effects of mass ratio of CMC, the molecular weight of PDMAEMA and PLA on the physical properties and thermal properties were investigated by Instron tensile machine and thermogravimetric analysis (TGA). The morphologies of the composites were characterized by SEM (Scanning Electronic Microscopy). The properties of final composites can be tuned by the molecular weight or mass of different components and PDMAEMA is acted as good lubricant between stiff CMC and PLA.

(234d) Characterization of Blue Maize and White Sorghum Flour Films Made By Thermoplastic Extrusion

Trujillo-de Santiago, G.,
Rojas-de Gante, C., Tecnologico de Monterrey
Garcia-Lara, S.,
Iannace, S.,
Di Maio, E.,
Originally presented on: 11/5/2013 9:45:00 - 10:10:00

Abstract

The feasibility to produce flexible films from blue maize (BM) and white sorghum (WS) flours plasticized with sorbitol/glycerol through extrusion/compression was evaluated in this work. Screws speed rotation (50, 60, 70 rpm), extrusion temperature (100, 110, 120, 130 and 140°C), number of extrusion passes (1 or 2) and chemical modification of flour with maleic anhydride were the analyzed variables. Flexible films were obtained from all the treatments with tensile strength values in the range of 0.7-2.3 MPa and strain at break in the range of 0.08-0.74 mm/mm. Mechanical tests results showed that one extrusion cycle at 120°C and 60 rpm were the best tested. Extrusion temperature had effect weather in mechanical properties, microstructure and color of films; moreover FTIR analysis showed the induction of oxidized products at higher temperatures. Compared with native flours, films made with chemically modified flours increased tensile strength in a 29% for WS films, and strain at break in a 517% in BM films. Consistently, more uniform microstructure was observed by SEM in films made with chemically modified flours, and DMA indicates better matrix-plasticizer interaction than films made with native flours. BM and WS flours are suitable raw materials to produce thermoplastic films, although effective strategies to get materials with better performances have to be developed.

(234e) Cellulose Nanofibers for Advanced Applications

Jiang, L., North Dakota State University
Xu, X.,
Originally presented on: 11/5/2013 10:10:00 - 10:35:00

Two types of cellulose nanofibers, i.e. cellulose nanowhiskers and cellulose fibrils, were used in preparing polymer nanocomposites, electrospun nanofibers, and core-shell carbon fibers. Their effects on solution rheology, composite properties, nanofiber properties and carbon fiber properties were systimatically studied and compared. The two different cellulose nanofibers showed different effects mainly due to their different surface properties and aspect ratios.

(277a) Biomass Gasification Using Carbon Dioxide: Chemistry and Process Modeling

Sadhwani, N., Auburn University
Eden, M. R., Auburn University
Adhikari, S., Auburn University
Originally presented on: 11/5/2013 8:30:00 - 8:55:00

Gasification is partial oxidation of biomass to produce syngas, which can be used as a feedstock in industrial applications for electricity, chemicals and fuel production. Air, oxygen and steam have been widely used as oxidizing medium in the gasification process. The use of Carbon dioxide (CO_2) as the oxidizing agent in biomass gasification has been explored in limited studies, and it is shown to improve CO production and biomass conversion. The focus of this study is to understand the chemistry or sequence of reactions occurring during the process.

The process of biomass gasification using CO_2 as the oxidizing agent is carried out in a bench scale bubbling fluidized bed reactor at different temperatures (600^0C - 850^0C). The syngas composition, tar concentration and biomass conversion are used to determine the dominant set of reactions occurring within the given range of temperature. A number of gas analysis instruments such as a GC/MS (Agilent 7890/5975), a micro GC (Agilent 490), a portable gas analyzer (Nova 7905AQN4), and an FTIR are used to collect necessary information.

Furthermore, a model for this process is developed using ASPEN plus. The model utilizes Gibbs free energy minimization to predict the effect of operating parameters such as temperature, pressure, and oxidizing media on biomass conversion and syngas composition. The results from the model for the given conditions are compared to the experimental results and possible reasons for deviations are proposed.

(277b) Engineering Next Generation Biomass Feedstocks: Effect of Torrefaction On Gasification Products

Jablonski, W., National Renewable Energy Laboratory
Carpenter, D., National Renewable Energy Laboratory
Olstad, J. L., National Renewable Energy Laboratory
Oddo, M.,
Black, S., National Renewable Energy Laboratory
Robichaud, D., National Renewable Energy Laboratory
Originally presented on: 11/5/2013 8:55:00 - 9:20:00

Growing energy demand worldwide requires a solution that is both domestic and stable, regardless of the region. Renewable energy sources have been implemented and proven in affluent nations in Europe and North America, but it remains that these technologies only satisfy a fraction of the energy requirement globally. Biomass-derived energy sources are abundant, carbonaceous, diverse and domestic. Low thermal efficiencies and uneconomical process operation have limited the implementation of these sources. However, research for cost reduction and process improvement over the last 30 years has demonstrated that economical biomass-derived energy is achievable. In this work, we gasified several traditional and advanced feedstocks at 800°C and 1 atm in a 10.2 cm i.d. gasifier with a continuous solid feeder, hot char removal, and a dodecane scrubber for vapor condensation and separation. Gas chromatography (GC), non-dispersive infrared spectroscopy (NDIR), and a diode laser were used to speciate and quantify non-condensable gases, hydrogen sulfide, and ammonia in the raw syngas stream. Molecular beam mass spectrometry (MBMS) was used to measure and quantify tar species in the hot vapor product stream (~425°C). Idaho National Laboratory (INL) provided all feedstocks including: torrefied "clean" and "whole" southern pine, non-pretreated clean and whole southern pine, oak, and switch grass. Bark, branches, and needles were removed from "clean" southern pine while "whole" pine comprises the whole tree. Torrefaction (thermal pretreatment) was performed at 270°C in a fluidized sand bed, and the resultant solid sample was sieved for feeding to the gasifier. Non-condensable gas quality (e.g. CO, H_2, CO_2, CH_4, C_2H_4) did not vary greatly between the feedstocks as has been shown in recent work in the literature.[1] The torrefied feedstocks had markedly lower tar concentrations: 17.4% for untreated whole pine and 10.2% for torrefied whole pine. Recent studies have demonstrated that torrefaction deacetylates hemicellulose in the biomass thereby breaking down a main plant constituent without degrading the majority of the biomass (> 90% mass retention).[2]Based on work in the literature and results from this study, torrefaction reduced the tar content by unraveling the hemicellulose and cellulose polymers in the native biomass. Additionally, oxygen reduction during torrefaction increased the energy density of the biomass feedstock (17 MJ/kg to 19.6 MJ/kg). This densification makes the feedstock easier to transport, grind and pelletize. Future work will focus on using thermal pretreatment to homogenize blended biomass feedstocks to increase the pool of feedstocks resources to include more waste materials.

1. Couhert, C.; Salvador, S.; Commandre, J.-M., Impact of torrefaction on syngas production from wood. *Fuel* **2009,** *88*, 2286-2290.

2. (a) Melkior, T.; Jacob, S.; Gerbaud, G.; Hediger, S.; Le Pape, L.; Bonnefois, L.; Bardet, M., NMR analysis of the transformation of wood constituents by torrefaction. *Fuel* **2012,** *92*, 271-280; (b) Tumuluru, J. S.; Sokhansanj, S.; Hess, J. R.; Wright, C. T.; Boardman, R. D., A review on biomass torrefaction process and product properties for energy applications. *Industrial Biotechnology* **2011,** *7* (5), 384-701.

(277c) Operations of a Stratified Downdraft Gasifier On MSW for Development of a Waste to Energy Process

Lucero, A., Southern Research Institute
West, J., Southern Research Institute
McCabe, K., Southern Research Institute
Gangwal, S., Southern Research Institute
Originally presented on: 11/5/2013 9:20:00 - 9:45:00

Southern Research is developing a small-scale MSW and biomass conversion process based on gasification to produce renewable power and liquid fuels. At the heart of the process is a simple, low cost gasifier that has been proven in several installations worldwide as scales of 1 to 50 tons/day and that can convert minimally processed feeds to syngas containing low concentrations of tar, particulates, and other contaminants. Southern is targeting development of the process at 3 scales: 1 to 3 tons per day for US DoD forward operating bases and other small generator, 10 to 50 tons/day for US DoD fixed installations, other institutional generators, and municipal landfills, and a 100 tons/day microrefinery for producing liquid fuels. This presentation will focus on the operations of a stratified downdraft gasifier on minimally prepared synthetic MSW.

(277d) Effects of Oxygen and Steam On Biomass Gasification and Syngas Combustion in a Pilot-Scale System

Huynh, C. V., Iowa State University
Kong, S. C., Iowa State University
Originally presented on: 11/5/2013 9:45:00 - 10:10:00

This study is to investigate the characteristics of a biomass gasification and syngas combustion system using mixtures of "oxygen-enriched air" and steam as the gasifying agent for increasing the syngas heating value. This study also aims to characterize the effects of oxygen-and-steam gasification on ammonia concentration that can lead to significant NOx emissions from syngas combustion. Experiments are conducted using a pilot-scale pressurized bubbling fluidized bed gasifier with a capacity of five tons per day, rated 800 kW thermal. Pure oxygen is added to air before mixing with steam for gasification. A significant amount of steam is required to control the reactivity of the system at high oxygen levels. The oxygen content in the enriched air varies from 21, 45, to 80 vol% on dry basis, corresponding to 21, 30, and 40 vol% on wet basis respectively. The bed temperature is maintained at 800 °C for all tests. Three different biomass feedstocks with nitrogen contents varying from 0.5 to 1.4 wt% are used for study (i.e., pine, maple-oak mixture, and discarded seed corn). The syngas dry composition is measured using a micro gas chromatograph while ammonia concentration and moisture content are measured using a modified IEA Tar Protocol and Karl Fischer Titration respectively. Results indicate that oxygen-enriched air and steam gasification favors the production of combustible gas components including hydrogen, carbon monoxide, methane, and lighter hydrocarbons. When 40% oxygen is used, hydrogen increases by 70%, 47%, and 32% for pine, maple-oak, and seed corn respectively, while CO increases by 34%, 18%, and 8.6% respectively. Overall, it is found that oxygen and steam gasification is most effective for feedstock with low nitrogen and moisture contents. Results also show that ammonia and NOx concentrations in syngas increase as oxygen enrichment increases. The lower heating value of syngas can increase by as much as 43% for the feedstock studied. When the oxygen level increases from 21% to 40%, the H_2/CO ratio also increases from 0.59 to 0.75, 0.67 to 0.84, and 0.36 to 0.43 for pine, maple-oak, and seed corn respectively. Despite the improvement, the H_2/CO ratio is still moderate. The moderate H_2/CO ratio is explained by the high water content in syngas at high oxygen and steam conditions, indicating a large amount of un-reacted steam at the current gasifier temperature at 800 °C.

NO_x emissions from combustion of syngas derived from the above gasifier were also measured. The biomass-derived syngas was burned in an industrial burner that was integrated into the gasification system. The burner is rated at 879 kW thermal. For each set of biomass-derived syngas, NO_x emissions were measured at different burner operating conditions including various heat rates and equivalence ratios using emission analyzers with chemiluminescence technology. All the combustion test conditions are in the lean mixture ranges in order to avoid the peak temperature limitation of both the burner and combustion chamber.

Results show that NO_x emissions using syngas obtained from woody feedstock decrease almost linearly as the combustion mixture becomes leaner and the heat rate decreases. When compared to natural gas, syngas from both woody feedstock generates higher NO_x emissions even when the heat rates are comparable, indicating that fuel NO_x formation is highly important in biomass-derived syngas combustion. In contrast to syngas from woody feedstock, syngas from seed corn results in peak NO_x emissions before NOx decreases with further leaner conditions. The trend is observed for all fuel flow rates and all oxygen-enriched-air and steam conditions of seed corn-derived syngas. Among the three feedstock, seed corn has the highest nitrogen content which yields the highest ammonia concentration in syngas, which, in turn, results in the highest NO_x emissions for all test conditions. Overall, the NO_x emissions from seed corn-derived syngas combustion are approximately in the range of 450 ~ 900 ppm higher compared to those from wood-derived syngas combustion.

(277e) Demonstration of Biomass Conversion to Mixed Alcohols At the Pilot Scale

Hensley, J. E., National Renewable Energy Laboratory
Jablonski, W., National Renewable Energy Laboratory
Smith, K., National Renewable Energy Laboratory
Bain, R. L., National Renewable Energy Laboratory
Schaidle, J. A., University of Michigan
Phillips, S., National Renewable Energy Laboratory
Nimlos, M. R., National Renewable Energy Laboratory
Magrini, K., National Renewable Energy Laboratory
Davis, M., National Renewable Energy Laboratory
Originally presented on: 11/5/2013 10:10:00 - 10:35:00

An integrated thermochemical biomass to ethanol process was demonstrated at the pilot scale. This process entailed both fuels synthesis catalyst activation and operation on biomass derived syngas. The sum of these activities comprised more than 500 hours of pilot scale operation. The fully integrated biomass to mixed alcohol process was comprised of a solids feeder, fluidized bed indirect steam gasifier, thermal cracker, char collector, fluidized bed steam reformer, packed bed polishing steam reformer, scrubber, pressure-swing CO_2 adsorber, and gas-phase continuously stirred tank gas-to-liquids reactor (CSTR). Tars and methane were reformed using NREL-developed steam methane reforming catalysts and mixed alcohols (including ethanol) were produced using a metal sulfide catalyst. The goals of this study were to identify operating conditions and syngas compositions that are obtainable with NREL's gasifiers and gas cleanup equipment, to prove an ability to operate continuously for long periods, and to identify major impacts that biomass syngas may have on the fuel synthesis catalyst as compared to bottled syngas, as well as demonstrate scaled liquids production.

During this talk, the performance of the entire integrated process and each process unit will be discussed, and the key lessons learned will be addressed. The implications of this pilot-scale demonstration on future research as well as on commercialization of mixed alcohol production processes will also be discussed.

(277f) Optimization of a 3ton/Day Biomass Fed Bubbling Fluidized Bed Gasification System

Buchireddy, P., University of Louisiana at Lafayette
Guillory, J., University of Louisiana at Lafayette
Zappi, M., University of Louisiana at Lafayette
Vutukuri, J., University of Louisiana at Lafayette
Originally presented on: 11/5/2013 10:35:00 - 11:00:00

Biomass can be converted to different forms of energy via several routes which can be mainly classified into two groups, a) Thermochemical conversion and b) Biological conversion. Thermochemical conversion can be further sub-classified into combustion, gasification, pyrolysis, liquefaction, etc. Conversion of biomass to energy by biological routes involves various techniques including fermentation, digestion, extraction etc. Biomass gasification is the most promising thermochemical route for converting biomass to energy. Gasification process involves partial oxidation of carbonaceous fuels at high temperatures to produce an energy carrier. Gasification of biomass produces fuel gases (producer gas or synthetic gas), which can be used in the generation of electricity, production of transportation fuels and chemicals, hydrogen fuel production etc.

University of Louisiana at Lafayette in collaboration with Cleco Power LLC. (Major Electric Utility Company in Louisiana), has designed and installed 3 ton/day bubbling fluidized bed gasification system to accommodate 1) suitability for biomass waste feedstock, 2) compact configuration with future semi-portability, 3) operate using both air and oxygen as oxidizing medium, and produce syngas usable to generate electricity and/or produce liquid transportation fuels. The system has been installed in 2012 and the reactor has been tested and operated successfully using pine as feedstock. The system is currently being optimized to improve the overall gasification efficiency and produce improved syngas both in terms of quality and energy density.

Tests are being conducted using air and oxygen as gasifying medium, and varying operational parameters such as equivalence ratio, residence times, etc. to optimize the gasification system. In addition, a tar and particulate sampling system has been installed according to the standard tar and particulate sampling guideline, which will be used to evaluate the quality of syngas produced. Overall performance of the pilot scale biomass gasification system will be presented, in addition to the results obtained while optimizing the gasification system. Also, a 25 KW low BTU genset has been installed to produce electricity from biomass, the results of which will be presented.

(291a) Tuning Interactions Between Cellulose Nanocrystals With Adsorbing Polysaccharides and Surfactants

Hu, Z.,
Cranston, E.,
Pelton, R., McMaster University
Originally presented on: 11/5/2013 12:30:00 - 12:55:00

Cellulose nanocrystals (CNCs) are strong, light-weight, renewable, and highly crystalline rod-like nanoparticles of cellulose. CNCs from wood or cotton are nanorods of approximately 200 nm in length and 6-10 nm in width and are hydrophilic due to the presence of surface hydroxyl groups and charged sulfate ester groups. Because CNCs have recently become available in commercial scale quantities, they are being evaluated in a wide range of potential application. For example, CNCs have been incorporated into renewable polymer matrices to create sustainable, nanocomposites and assembled into optically active films that reflect specific wavelengths of light. We anticipate that because of the green, biocompatible nature of CNCs that they will be considered in formulated chemical products such as cosmetics. For these types of applications it is important to understand the influences of water-soluble polymers and surfactants on the colloidal stability and rheological properties of CNCs suspensions.

The aim of this paper is to investigate the structure formation and rheological properties of suspensions of in the presence of polymers and surfactants. Adsorption of hydroxyethyl cellulose (HEC), hydroxypropyl guar (HPG), locust bean gum (LBG), and dextran (Dex) onto cellulose nanocrystals- coated substrates was investigated. Measurements using the quartz crystal microbalance with dissipation (QCM-D) confirmed that all polysaccharides except dextran adsorbed onto cellulose nanocrystals and formed viscoelastic gel layers with water contents over 90%. Shear-thinning hydrogels were readily produced by mixing CNCs with adsorbing polysaccharides over the concentration range studied. Viscoelasticity measurements showed a larger linear elastic moduli than loss moduli, confirming gel formation. These results lead us to conclude that adsorbing polymer-induced bridging flocculation of cellulose nanocrystals causes gel formation. The stability of CNCs-polymer hydrogels was tailored by exposing the hydrogels to sodium dodecyl sulfate solutions. Complementary use of QCM-D and surface plasmon resonance (SPR) spectroscopy indicated that polymers desorb from the cellulose surface in the presence of sodium dodecyl sulfate. A depletion flocculation mechanism was suggested to explain the decrease in viscosity for cellulose nanocrystals-polymer hydrogels upon addition of anionic surfactants. We believe that the more detailed understanding of the network formation and rheology of CNCs suspensions in the presence of polymers and surfactants will aid in incorporating CNCs into various formulations extending the range of possible applications.

(291b) Fabrication of Electronic Devices Using Highly Flexible Microfibrillated Cellulose

Zhang, X., Georgia Institute of Technology
Deng, Y., Georgia Institute of Technology
Sharma, S., Georgia Institute of Technology
Lin, Z.,
Chen, B.,
Originally presented on: 11/5/2013 12:55:00 - 13:20:00

Cellulose, as the most abundant biopolymer in the world and the main component of papers, is a polysaccharide consisting of thousands of β -linked glucose unit. The hydroxyl groups on the backbone of cellulose make it easy to be functionalized. In the meanwhile, modern technology requires electronic devices to be more flexible and environmental friendly, which makes cellulose to be a candidate to fabricate electronic devices. Up to now, lots of researches employ "paper-like" plastic polymers to fabricate electronics rather than using real cellulose paper. Cellulose, compared to its counterparts, is much cheaper, more environmental-friendly and has the excellent properties of being flexible and mechanically strong. There are also some general challenges about using cellulose for electronics, such as its nonconductivity, porosity and roughness, but these features can be taken advantages of on some occasions. In my study, two different cellulose based electronic devices will be studied. For the first part, an ionic paper-based diode(IPD) was made by combining together two oppositely charged microfibrillated cellulose sublayers. Unlike the conventional p-n junctions using electrons and holes as charge carriers, the IPD relies on transportation of ions to conduct current. The asymmetric charge distribution between the two sublayers will help selectively transport cations and anions under positive and negative bias, to allow an electric current to pass in only one direction. For the second part, flexible and free-standing supercapacitors made mainly by muti-walled carbon nanotubes(CNTs) and micro-fabrillated cellulose(MFC) was investigated. The results indicated that the as-made supercapacitor showed a high specific capacitance of 176mF/cm^2. In addition, the electrochemical properties of the supercapacitors remained almost the same under severe bending. In sum, cellulose-based electronics have good characteristics of being low-cost, green and flexible and can be applied in various areas, such as supercapacitors, transistor, solar cells, etc.

(291d) New Capabilities for Processing Biomaterials With a Tandem Micro-Reactor GC/MS System for Catalysis Research

Ramus, T.,
Ramus, T.,
Ramus, T.,
Originally presented on: 11/5/2013 13:20:00 - 13:45:00

The tandem micro-reactor is designed for rapid evaluation and characterization of catalyst systems. Three modes of operation allow different experiments to be conducted on the same equipment. The GC/MS can operate in an on-line MS mode for continuous analysis of gases from the catalyst bed. Alternatively, the GC/MS can operate in several integrated high resolution GC/MS modes for step-wise experiments on gases from the catalyst bed. A single-shot mode allows the analysis of a gas, liquid or solid sample. The system consists of an upper micro-furnace and lower micro-reactor each with independent temperature and reaction gas control. The micro-furnace can accept solid or viscous liquid samples in a batch sampling mode with catalysis occurring under different conditions in the micro-reactor. The micro-reactor is designed to allow a quick change of the catalyst bed. Batch or continuous experiments can be performed with this system to evaluate both catalyst performance and to characterize catalysis products. Provisions are also designed into the system to allow catalyst regeneration and evaluation during this step. The applications include processing cellulose materials.

(291e) Synthesis and Curing of Phenol- 5-(Hydroxymethyl)Furfural (PHMF) Novolac Resin and Its Formaldehyde-Free Curing With Organosolv/Kraft Lignin (OSL/KL)

Zhang, Y., Institute for Chemicals and Fuels from Alternative Resources (ICFAR)
Yuan, Z., Institute for Chemicals and Fuels from Alternative Resources (ICFAR)
Xu, C., Lakehead University
Originally presented on: 11/5/2013 13:45:00 - 14:10:00

The use of formaldehyde to prepare phenol-formaldehyde (PF) resins is one of the primary challenges for the world-wide PF industry with respect to both sustainability and human being health. Lignocellulosic biomass has tremendous potential as a feedstock to make renewable polymers. This study reports the synthesis of novel glucose-based PHMF resin and its curing reaction with OSL/KL to produce formaldehyde-free phenolic resins. The use of HMF which was converted from glucose allows an economical and green route to synthesize phenolic resin without formaldehyde consumed or released and enhances its reactivity towards phenol. The chemical structure of PHMF was characterized by Fourier transform infrared (FTIR) spectroscopy and proton nuclear magnetic resonance (^{1}H NMR). Polymer molecular weight was determined by Gel permeation chromatography (GPC). OSL/KL, which are rich in woody biomass, were proven to be environmentally benign curing agents for PHMF. The curing mechanism was studied by carbon NMR (^{13}C NMR), which demonstrated the occurrence of alkylation reaction between lignin's hydroxylmethylene group and PHMF's para- carbon of phenolic hydroxyl group. Further studies regarding curing kinetics were conducted by using dynamic differential scanning calorimetry (DSC) and isoconversional methods to find out kinetic parameters. For the purpose of its application as replacement of PF resin, performances of OSL/KL cured PHMF were tested in terms of thermal stability, and mechanical properties. This research project realized curing of phenolic resin with OSL/KL for the first time among open publications. The results obtained presented OSL/KL to be promising curing agents for PHMF and recent results presented here will help make eco-friendly phenolic adhesive a practical reality.

(330a) Adsorption of the Compounds Encountered in Monosaccharide Dehydration in Zeolites

Leon, M., University of Delaware
Xiong, R., University of Delaware
Swift, T. D., University of Delaware
Nikolakis, V., University of Delaware
Sandler, S. I., University of Delaware
Vlachos, D. G., University of Delaware
Originally presented on: 11/5/2013 12:30:00 - 12:52:00

Biomass carbohydrates are the most abundant renewable resources available and have the potential to replace petroleum as a source of both energy and chemicals, leading to more environmentally friendly processes and reducing the dependence on crude oil. A promising reaction in this field is the acid-catalyzed dehydration of monosaccharides, such as fructose and glucose, towards 5-hydroxymethylfurfural (HMF), an important intermediate for the production of valuable chemicals.[1] When zeolites are used as catalysts the preferential adsorption of reactants, products or byproducts could markedly affect the yield to the desired product. In this regard, studying the adsorption of the compounds present in the reaction media in the zeolite catalyst becomes crucial to understand the reactivity of the system and it is necessary to calculate the intrinsic reaction rate constants from kinetic data. Zeolites can also be used as adsorbents for the selective removal of the HMF from the reaction media (either in-situ or after dehydration). In this respect understanding the effect of the framework type, composition (e.g. Si/Al, extraframework cation), and presence of co-solvent (e.g. dimethylsulfoxide (DMSO)) on HMF adsorption are important for the design of an adsorptive separation process.

Adsorption loadings from solution are usually estimated from changes of the solute concentration assuming that the solution volume is not affected by adsorption. This quantity is called excess adsorption loading and sometimes it is quite different from the real loading. We formulated and used a method for the estimation of the real adsorption loading from the experimentally measured excess adsorption that can be applied to calculate the adsorption isotherms both in the case of single-solute and multi-solute mixtures.

The effect of temperature on the adsorption of fructose, glucose, HMF, levulinic and formic acid from aqueous solutions in the protonated form of zeolite beta with SiO_2/Al_2O_3 ratio of 18 (H-BEA) will initially be presented. The limiting heat of adsorption at zero coverage of all components was calculated from isotherms measured at 0, 25 and 40 °C. Adsorption experiments from multi-solute solutions were also carried out and the data were found to be in good agreement with predictions made using Ideal Adsorbed Solution Theory (IAST) and the single solute isotherm experimental data.

Adsorption of HMF was studied in more detail. We examined the effect of the framework type (BEA, MFI, FAU), composition (Si/Al) as well as the addition of DMSO as co-solvent on its adsorption uptake. It was found that in each zeolite type the adsorption uptake of HMF increases with the hydrophobicity (higher Si/Al ratio) of the zeolite. Experimentally measured isotherms were in good agreement with those calculated using Grand Canonical Monte Carlo - Expanded Ensemble (GCMC-EE) simulations. It is found that in the case of HMF/water binary mixtures, zeolites with larger pores (i.e. FAU), have higher HMF adsorption capacity but lower selectivity because large pores allow water adsorption due to favorable H_2O-HMF attraction. Therefore, dealuminated zeolite Y is less selective for the separation of HMF from water than silicalite-1 (siliceous MFI type zeolite) that has smaller pores sizes. In addition, experiments and simulations show that in HMF/DMSO/water ternary mixtures, the HMF loading decreases with increasing DMSO concentration. This decrease is due to the high solubility of HMF in DMSO, and to the fact that in the entropic contribution to the free energy of adsorption dominates over the enthalpic contribution. Analysis of our data shows that the ratio of the energetic interaction in the zeolite to the solvation free energy is a key factor in controlling separation from liquid mixtures.

To conclude, our findings could shed light on the acid catalyzed dehydration of monossacharides and the HMF separation, and serve as a hint to the rational design of microporous adsorbents for liquid-phase separations in biomass processing.

REFERENCES

1. Chheda, J. N.; Huber, G. W.; Dumesic, J. A. Liquid-Phase Catalytic Processing of Biomass-Derived Oxygenated Hydrocarbons to Fuels and Chemicals. *Angew. Chem. Int. Ed.* **2007,** *46*, 7164-7183.

2. Vandam, H. E.; Kieboom, A. P. G.; Vanbekkum, H. The Conversion of Fructose and Glucose in Acidic Media - Formation of Hydroxymethylfurfural. *Starch-Starke* **1986,** *38*, 95-101.

3. Kruger, J. S.; Nikolakis, V.; Vlachos, D. G. Carbohydrate Dehydration Using Porous Catalysts. *Curr. Opin. Chem. Eng.* **2012,** *1*, 312-320.

4. Kruger, J. S.; Choudhary, V.; Nikolakis, V.; Vlachos, D. G. Elucidating the Roles of Zeolite H-Bea in Aqueous-Phase Fructose Dehydration and HMF Rehydration. *ACS Catal.* **2013**.

5. Roman-Leshkov, Y.; Chheda, J. N.; Dumesic, J. A. Phase Modifiers Promote Efficient Production of Hydroxymethylfurfural from Fructose. *Science* **2006,** *312*, 1933-1937.

(330b) Elucidating Structure-Properties Relations for the Design of Highly Selective Carbon-Based HMF Sorbents

Yoo, W. C.,
Rajabbeigi, N., University of Minnesota
Mallon, E. E., University of Minnesota
Tsapatsis, M., University of Minnesota
Snyder, M. A., Lehigh University
Originally presented on: 11/5/2013 12:52:00 - 13:14:00

Continuously increasing energy demands in a fragile global market and in the face of growing environmental concerns motivate the drive to develop sustainable energy resources like plant biomass for replacing or augmenting current petroleum-derived fuels and chemicals [1-3]. One of numerous challenges facing the biorefinery is the selective and efficient production of important chemical products such as 5-hydroxymethyl furfural (HMF). HMF production from fructose dehydration in DMSO with various catalytic systems has shown superior selectivity and high yields [3, 4]. Yet, the subsequent need for energy-intensive separation of HMF from HMF/fructose/DMSO mixtures challenges the practical feasibility of this process. Microporous carbon materials have shown promise as substrates for alternative adsorption-based separations [5], but the origin of HMF selectivity and adsorption capacity, and thereby structure-properties relations enabling rational design of enhanced sorbents, has remained elusive. Through systematic quantification of the functionality and texture of various commercial (BP2000, Norit1240) and synthetic carbons (three-dimensionally ordered macroporous carbons; mesoporous carbons), we link, for the first time, adsorption capacity and HMF adsorption selectivity to two tunable materials properties of the carbon sorbents: micropore texture and oxygenate functionality (i.e., carbon polarity). In the process, we exploit these newly elucidated structure-properties relations for realizing a class of synthetic carbons with sub-micron particulate morphology that achieve more than 60% higher selectivity and more than 20% higher capacity for HMF over fructose as compared to the best performing commercial product, BP2000.

References

[1] P.Y. Dapsens, C. Mondelli, J. Perez-Ramírez, ACS Catal., 2 (2012) 1487-1499.

[2] G.W. Huber, S. Iborra, A. Corma, Chem. Rev., 106 (2006) 4044-4098.

[3] A. Corma, S. Iborra, A. Velty, Chem. Rev., 107 (2007) 2411-2502.

[4] X. Qi, M. Watanabe, T.M. Aida, R.L. Smith Jr, Ind. Eng. Chem. Res., 47 (2008) 9234-9239.

[5] N. Rajabbeigi, R. Ranjan, M. Tsapatsis, Microporous Mesoporous Mater., 158 (2012) 253-256.

(330c) Separation of Lignin From Hydrolyzates By Flocculation Using PEO Polymers

Ramarao, B. V., ESPRI, SUNY College of Environmental Science and Forestry
Yasarla, R., State University of New York
Gnanavel, V., SUNY ESF
Originally presented on: 11/5/2013 13:14:00 - 13:36:00

Hydrolyzates produced by pretreatment of lignocellulosic materials contain phenolic compounds, organic acids and aldehydes such as furfural in colloidal and dissolved form. Since these compounds can be potent inhibitors for fermentation, they need to be separated and isolated ahead of such biological treatments. We found that the application of a strongly binding non-ionic polymer, polyethylene oxide causes flocculation and rapid separation of lignin from the hydrolyzates, while preserving the carbohydrates in solution to a large extent. The effect of flocculation is measured by the reduction in turbidity and enhancement of the suspension settling velocity. The maximum flocculation was found to depend on the dosage of the polymer and its MW, but was independent of the pH of the hydrolyzates (from 2 through 9). PEO dosages were between 20 to 50 ppm at the maximum flocculation and reduction in suspension turbidity. The suspension temperature plays an important role, with temperatures below 21.5 C showing no flocculation.

The hemicellulose fraction in the supernatants was close to the raw extract showing that flocculation does not remove significant amount of fermentable sugars. Hence such pretreatment clarification is not likely to affect downstream bioproduct yields. Flocculation and clarification with PEO were also demonstrated on large pilot scale batches of 1000 kg of extracts. The flocculated extract was filtered and the filtrate showed a 99.5% reduction in turbidity, from 12000 NTUs in the raw extract to 50 NTUs in the filtrate. The impact of flocculation on solid liquid separations was quantified and flocculation was optimized with respect to dosage, suspension concentration, temperature and pH.

(330d) Fractionation and Recovery of Lignin and Xylan From Alkali Liquors - Characterization for Biorefinery Processes

Stoklosa, R. J., Michigan State University
Hodge, D., Michigan State University
Originally presented on: 11/5/2013 13:36:00 - 13:58:00

Understanding the behavior of plant cell wall biopolymers during alkaline biorefining processes and how their properties are affected by the processing conditions is important for maximizing value from woody lignocellulosic feedstocks. In this work, the fractionation of alkali pulping liquors was investigated to characterize the extractability, recoverability, and properties of the solubilized hemicelluloses and lignin during the alkali impregnation and heat-up stages of alkali organosolv pulping or soda pulping for three hardwood species: sugar maple (*Acer saccharum*), silver birch (*Betula pendula*), and a hybrid poplar (*Populus nigra x maximowiczii cv.* NM6). Recoverability of the solubilized biopolymers were determined using a variety of methods including aggregation and precipitation by acidification as well as ultrafiltration. The xylan showed a decreasing degree of polymerization (DP) with increasing temperature during reactor heat-up, which stabilized at a DP between 10-30. Lignin aliphatic and aromatic hydroxyl content was characterized by ^{1}H NMR and the lignin molecular weight distributions were determined by gel permeation chromatography (GPC). These properties showed a clear dependence on the extraction conditions and were clearly linked to recoverability of lignin by either acidification or ultrafiltration.

(330e) Novel Catalytic Membranes for Combined Biomass Hydrolysis and Sugar Recovery

Avram, A., University of Arkansas
Lei, J., University of Arkansas
Qian, X., University of Arkansas
Wickramasinghe, S. R., University of Arkansas
Beier, A., University of Arkansas
Originally presented on: 11/5/2013 13:58:00 - 14:20:00

Production of biofuels from lignocellulosic biomass is one of the leading candidates for replacement of fossil fuels. In the biochemical conversion platform, dilute sulfuric acid pretreatment is one of the leading technologies to hydrolyze hemicellulose, relocate lignin and expose cellulose. After pretreatment, a cocktail of cellulase enzymes is used to break down cellulose synergistically. The major problems associated with this technology are the slow and incomplete hydrolysis of cellulose and the degradation of monomer sugars during pretreatment. Furthermore, the cost of the enzymes has been an inhibitory factor for the commercialization of biomass conversion technology.

Here, a novel catalytic membrane reactor is designed to simultaneously hydrolyze cellulose and hemicellulose and separate the hydrolyzed sugars, thus minimizing product degradation. A polymeric solid acid catalyst is synthesized to replace the corrosive sulfuric acid and expensive enzymes, but operates at a higher conversion rate than the analogous natural enzymes and that costs far less to produce. We have grown adjacent nanostructure consisting of tethered polysulfonic acid and polyionic liquid chains from the surface of ceramic membranes. The feed consisting of cellulose and hemicellulose in a water / ionic liquid mixture is introduced on one side of the membrane. The polyionic liquid chains help solubilize the polysaccharides while the polysulfonic acid chains catalyze the hydrolysis of the polysaccharides. The membrane pore size is such that only water containing dissolved monomeric sugars and short chain oligomers can pass through. The ionic liquid and larger polysaccharides are rejected.

From preliminary hydrolysis experiments, our results indicate the potential of these novel membranes to hydrolyze polysaccharides, minimize degradation product and save on catalyst costs.

(330f) Optimization of An Extractive Distillation With Mixed Solvents As Entrainer

Garcia, L. C., Universidad Nacional de Colombia
Gil, I. D., Universidad Nacional de Colombia
Orjuela, A., Universidad Nacional de Colombia
Rodriguez, G., Universidad Nacional de Colombia
Originally presented on: 11/5/2013 14:20:00 - 14:42:00

Anhydrous ethanol is widely used in the biofuels industry, since it has demonstrated to be environmental friendly and to have excellent properties for clean-burning. In mixtures of ethanol and gasoline, alcohol must have purity higher than 99.5% in order to avoid phase separation. Among the most popular processes used in ethanol dehydration, extractive distillation has demonstrated to be an adequate alternative due to the low energy consumption compared with other techniques and its economic feasibility. In this work, an extractive distillation process was simulated with the Aspen Plus software platform, using the RadFrac module for distillation columns. A mixture of ethylene glycol and glycerol is proposed as entrainer. This enhanced entrainer combines the low viscosity and the ability of the ethylene glycol to break the binary azeotrope with the low prices of glycerol. Phase equilibrium of this quaternary system was calculated with the NRTL model, which correlates well these strongly polar mixtures. The nominal case consists in the dehydration column, the recovery column and the solvent recycle loop. In order to establish the most adequate operating conditions, it was studied for each column the effect of the molar reflux ratio, the feed stage and the feed condition in the ethanol and solvent compositions. For the dehydration column, it was also investigated the effects of the ethylene glycol-glycerol mixture composition, the separating agent feed stages, the separating agent split stream, and the azeotropic feed temperature on the separation. Subsequently, the thermal and hydraulic analysis of both columns was established, so that the system operates energetically efficient. Finally, the economic optimization of the system was achieved, using the maximization of the profit as objective function. In this case, the solvent to feed ratio, the reflux ratio and the number of stages in the extractive column were the optimization variables, while the design parameters of the recovery column were fixed.

(330g) Study of Physical Properties of Binary Mixtures of Sulfolane + 2-Butanol (or + 2-propanol) At Different Temperatures

Mesquita, F., Federal University of Ceara
Sant'ana, H. B., Universidade Federal do Ceara
Feitosa, F., Universidade Federal do Ceara
Aznar, M.,
Santiago-Aguiar, R. S., Universidade Federal do Ceara
Originally presented on: 11/5/2013 14:42:00 - 15:04:00

Sulfolane and alcohols are classes of chemical compounds that are usually used as a solvent in different industrial recovery processes. The focus of the present work is to measure volumetric (densities) and transport (dynamic viscosity) properties of the following binary mixtures: sulfolane + 2-butanol (or + 2-propanol) at a range temperature from T = 303.15 to T = 343.15 K, at atmospheric pressure. Densities and viscosities were measured using a viscodensimeter (Anton Paar SVM 3000 Digital Oscillation U-tube). Viscosity deviations and excess volumes were obtained from the experimental data, by using a fitting procedure (Redlich-Kister type equation). The results show that excess volumes were positive over the entire mass fraction range for all binaries mixtures studied. Otherwise deviation viscosities are negative over the entire mass fraction range of the mixture. Excess volumes and deviation viscosities behaviors increasing with temperature.

(346a) Completely Biobased Lignin Adhesives for Plywood Applications

Li, R.,
Chung, Y. L.,
Sattely, E., Stanford University
Billington, S., Stanford University
Originally presented on: 11/5/2013 15:15:00 - 15:40:00

Lignin is one of the most abundant nonedible biomass found in nature and has the great potential for use as renewable feedstock for bio-based materials. When found in nature, lignin serves as a *bioglue*that covalently bonds and crosslinks various plant polysaccharides, affording plants mechanical strength. An application that directly draws on inspiration from lignin's biological role is as wood adhesives. Lignin's inherent low reactivity, however, is often only overcome by the use of toxic and environmentally harmful chemicals (e.g. formaldehyde, urea, or strong acids/bases) and conditions that are not amenable for use at the industrial scale.

In this study, we describe a simple and environmentally benign procedure for the production of lignin-based adhesive for plywood applications. Adhesion is achieved via direct cross-linking with bio-derived cross-linkers without the need for pre-activation of lignin. Adhesion can also be achieved via catalytic repolymerization of lignin under aerobic conditions. The resulting formulation has high lignin content (up to 60%) with overall total bio-derived feedstock content of up to 100%. The use of various lignin sources in the formulation was explored. Mechanical properties of plywood assembled with our lignin-adhesives were comparable with those assembled using commercial plywood adhesives.

The resulting bio-based adhesive formulation has high lignin content and contains only environmentally friendly and non-toxic components. Together with its ease of use in plywood assembly this formulation is potentially suitable for large-scale industrial applications.

(346b) Lignin As Reinforcing Filler for Elastomeric Compounds

Bahl, K.,
Jana, S. C., University of Akron
Originally presented on: 11/5/2013 15:40:00 - 16:05:00

Lignins possess strong mechanical properties and a number of reactive functional groups which qualify them as reinforcing fillers of polymer compounds. However, their polar nature makes them incompatible with non-polar rubber and their large particle sizes provide poor mechanical reinforcement. In addition, lignins are known to retard curing of rubber compounds. This study deals with how the functional groups on the lignin surfaces can be used effectively to solve the very issues that they create. The first part of this study focuses on modification of lignin with cyclohexylamine (CA) to derive compatibility with the rubber matrix. CA reduces the polarity of lignin via proton transfer and hydrogen bonding interactions. The second part of this study deals with solving the issues associated with large size of lignin particles. The non-covalent interactions between lignin and carbon black are used to coat carbon black particles. In this manner, the particle size of lignins is reduced in conjunction with a reduction of carbon-carbon particle networks. Both these effects lead to lower viscous dissipation and higher energy efficiency in rubber compounds. The effects of such hybrid fillers on tensile and dynamic mechanical properties of styrene butadiene rubber compounds are discussed

(346c) Synthesis and Curing of Bio-Based Epoxy Resin Derived From Organosolv Lignin

Ferdosian, F., Institute for Chemicals and Fuels from Alternative Resources (ICFAR)
Yuan, Z., Institute for Chemicals and Fuels from Alternative Resources (ICFAR)
Anderson, M.,
Xu, C., Lakehead University
Originally presented on: 11/5/2013 16:05:00 - 16:30:00

Epoxidation of lignin is a promising way to promote its performance and develop its new application in coatings and adhesive industries. In the present work, bio-based epoxy resin was synthesized by the reaction of de-polymerized organosolv lignin with epichlorohydrin in alkaline condition and in presence of a phase transfer catalyst. The effects of reaction time, reaction temperature and sodium hydroxide/lignin ratio were investigated. A statistical optimization process was also conducted to achieve a bio-based epoxy resin with high epoxy index and low molecular weight. The obtained bio-based epoxy resins were characterized using FTIR to investigate the chemical structure of lignin and confirm the grafting of epoxy groups on the structure on lignin and gel permeation chromatography (GPC) for their molecular weights and dispersion. The curing kinetic of the synthesized epoxy resin/ Diethylenetriamine (DETA) was studied by non-isothermal differential scanning calorimetry (DSC). The kinetics of the curing reaction was evaluated by the advanced isoconversional method which is proposed by Vyazovkin to obtain the activation energies of the curing reaction as a function of the degree of cure.

(346d) The Effect of Irradiation On the Processability of Lignin for Carbon Fiber Applications

Ayoub, A.,
Venditti, R., North Carolina State University
Originally presented on: 11/5/2013 16:30:00 - 16:55:00

Lignin is one of the most abundant macromolecules existing in the plant kingdom and the second major component of biomass. Lignin has gained interest as a source of fuel and new materials due to its renewability and its enormous amount available as a byproduct of the pulp and paper industry. Utilization of technical lignin as high value added applications is limited currently. There are several factors restricting the use of lignin namely: a non-uniform structure unique chemical reactivity, and the presence of various organic and inorganic impurities. One main disadvantage is that when heated to form a liquid the lignin degrades.

One of the most interesting value-added products that can be made from lignin is carbon fiber for composite or adsorbent applications. Fibers for high performance composites are one of the most important advanced engineering materials produced today. Lightweight fatigue resistance materials possess high strength, high modulus and high stiffness. Carbon fiber composite products are routinely used in sports equipment, marine products, construction, aircraft and the automotive industry. The conversion of lignin into valuable products is one way to add value to an existing wood processing process, an enabling concept of the biorefinery.

Due to its technical and environmental advantages over traditional solvent-based treatments, the processing of polymers under ionizing radiations has known increasing industrial developments for the past 20 years, in particular in the field of synthetic coating cross-linking. The radiation based modification of biopolymers is also a feasible method either to produce material or to tailor their usage properties. From a molecular point of view, irradiation was shown to generate free radicals on polymers and then induce various reactions as scission or crosslinking depending upon both the environmental conditions and the radiation dosage.

The objective of this research was to develop from lignin a stable thermoplastic material with a reasonable temperature window for processing in the melt state.

To meet the objective, softwood kraft lignin was modified using irradiation at low doses with and without styrene present and characterized. Irradiation of kraft lignin alone with radiation shows an initial decrease in molecular weight due to chain scission up to about 10kGy followed by an increase in molecular weight due to crosslinking. NMR results indicate a decrease of about 15% in the OH content of the lignin with increased irradiation after 10 kGy. Thermal properties such as Tg, free volume and Delta Cp follow accordingly. Irradiation at very low dosages was determined to facilitate the grafting of styrene monomer to lignin, decreasing the OH content by 23%. This effect increased the hydrophobicity of the material, depressed the value of Tg, increased the Delta Cp, increased the mobility in the liquid state, and made the material more thermally stable relative to the lignin alone, thus improving its processability at high temperatures. Both the irradiation of lignin alone and the grafting of styrene to lignin increased the yield of mass during pyrolysis and the activation energy for mass loss relative to untreated lignin alone. This work has demonstrated that the application of low dosages of irradiation is a promising method to attach functional molecules onto lignin for use in various applications.

This type of research is one of the first steps in modifying renewable lignin so that it is a more consistent and reliable precursor for material applications. Irradiation is a promising green tool in this respect and optimized/alternative methods to utilize irradiation to improve lignin as a precursor in a cost effective manner should be pursued.

(346e) Catalytic Hydrodeoxygenation of Lignin Derived Bio-Oils

Rahimpour, M. R., Shiraz University
Originally presented on: 11/5/2013 16:55:00 - 17:20:00

Bio-oils derived from biomass offer the prospect of becoming a major feedstock for production of fuels and chemicals, and lignin is a plentiful, underutilized source. Lignin conversion requires depolymerization and removal of oxygen. Likely processes for lignin conversion involve depolymerization (e.g., by pyrolysis) and catalytic upgrading of the resultant bio-oils. A major goal of the upgrading is catalytic hydrodeoxygenation (HDO), which involves reactions with hydrogen that produce hydrocarbons and water. The aim of this review is to present a critical introduction to HDO chemistry focused on compounds derived from lignin. This article provides a summary of the reactions of HDO and those that accompany it, with a comparison of catalysts and a discussion of their stabilities. The reactions are evaluated in terms of reaction pathways of compounds representative of lignin-derived bio-oils, including anisole, guaiacol, and phenol. The review includes recommendations for further research and an attempt to place HDO in a context of options for renewable fuels and chemicals, but it does not provide an economic assessment.

(407a) Manufacture of Value-Added Products By Ozonolysis From Soybean Oil

Zhao, Y.,
Graiver, D., Michigan State University
Narayan, R., Michigan State University
Originally presented on: 11/6/2013 8:30:00 - 8:55:00

The depletion of non-renewable resources by modern societies and the need for a sustainable chemical industry are major factors that contribute to an increasing interest to study and develop new biobased polymers. The generation of biobased building blocks for polymer synthesis is a primary research topic towards this goal. Soybean oil is a widely available resource, relatively cheap and contains the structure of unsaturated fatty acids, which can be modified into polymer building blocks by either (1) functionalization of these unsaturated fatty acids or (2) cutting them into small molecules while adding useful functional groups (e.g. carboxyl, hydroxyl, formyl, etc.). In this project, we will focus on the latter strategy by breaking the double bonds present in unsaturated fatty acids via ozonolysis. Ozone is known to be a very strong oxidant, capable of reacting with alkenes and breaking double bounds to generate formyl and carboxyl groups.

Advantages of using ozonolysis for oxidizing fatty acid double bonds include fast reaction rates and the low footprint of this chemical in the final product mixture. However, there are also some disadvantages that include the formation of a wide range of by-products, some of which are unstable compounds (e.g. ozonides). Moreover, ozonolysis reactions occur in a gas-liquid interface, which become hard to study and to measure the reaction kinetics.

In this work we will evaluate two solvent systems to perform ozonolysis reactions. In one case we utilized methanol as a solvent system to generate methyl-monoesters and -diesters, which can be utilized as polymer building blocks (e.g. polyesters). In a second case, we will utilize water as the ozonolysis solvent which allows the generation of aldehydes with yields ranging 75-77% in the best scenario. Some of these aldehydes (e.g. hexanal and nonanal) can be utilized as bio-based polymer building blocks. The kinetics of ozonolysis reactions will be studied as a function of reaction temperature, time and reactants concentration. A kinetic model will be elaborated as a function of these variables, and kinetic parameters will be determined by fitting the model equation to our experimental data through regression.

Purified aldehydes from the aqueous ozonolysis system, such as hexanal and nonanal, will be utilized in the synthesis of new biobased acetal polymers. Structural and performance characterization of these new biobased polymers will be performed to evaluate their potential areas of utilization and respective target markets.

(407b) Wheat Straw to Bio-Oil and Hydrogen Production From Aqueous Phase of Bio-Oil

Pouralamdari Meibod, M.,
Abedi, J., University of Calgary
Originally presented on: 11/6/2013 8:55:00 - 9:20:00

Currently, a viable route towards hydrogen production from biomass—such as agricultural residues— has been developed through the process of catalytic steam reforming of aqueous fraction of biomass pyrolysis oil. This method consists of two stages: generating bio-oil from biomass via fast pyrolysis reaction in the first stage, and thereupon applying the catalytic steam reforming process on the aqueous phase of bio-oil in order to produce renewable hydrogen in the second stage. The overall goal of this study is to evaluate both stages of this method and consequently the conversion process of producing renewable hydrogen from biomass. In the first stage (pyrolysis process), wheat straw as biomass was converted into pyrolytic oil, char and gases in a fluidized bed reactor. Under the optimum conditions of pyrolysis temperature of about 500 ^{o}C and wheat straw particle size less than 0.6 mm, the maximum achieved bio-oil yielded over 50 wt%. Furthermore and in the second stage (catalytic steam reforming process), the aqueous phase of a commercial bio-oil as well as the aqueous phase of bio-oil produced in the first stage were converted to hydrogen in a catalytic fixed bed steam reformer. For the case of commercial bio-oil, the effects of three groups of home-made Nickel-Alumina based catalysts as well as operational conditions of the process were evaluated in terms of their yield for hydrogen production. The highest hydrogen yield of about 65 % was achieved over the most effective catalyst (Ru-Ni/Al_2O_3 with 14 wt% Ni) in the under the following optimum conditions: temperature of 850 ^{o}C and a space velocity (WbHSv) of 10 h^{-1}. For the case of aqueous homemade bio-oil, hydrogen was produced with the yield of about 60% over the same catalyst and under similar operational conditions.

(407c) Only the Strong Survive: Saccharification Residues and Their Qualities As Lignocellulose By-Products

Schilling, J. S., University of Minnesota
Tze, W. T. Y., University of Minnesota
Yang, H. S., University of Minnesota
Liew, F. J.,
Hafez, I.,
Originally presented on: 11/6/2013 9:20:00 - 9:45:00

The biochemical platform for production of biofuels has progressed significantly in terms of conversion efficiency and net yields, but a portion of lignocellulosic plant tissue inevitably resists conversion. This residue portion, by the nature of its own recalcitrance, has qualities that differ from the starting material. Specifically, the enzymatic hydrolysis (saccharification) and dissolution of carbohydrate fractions leaves behind cellulose that is both highly crystalline and naturally well-dispersed in a lignin matrix. While often seen as an inefficiency, we are adopting the view that this is useful pre-processing amenable to making solid materials. As post-saccharification co-products from aspen and poplar, we are testing the potential to make lignin-based binderless films and adhesives reinforced in situ by cellulose nanocrystals. The work presented here relates to our objectives to examine pretreatment effects on material residue properties, including carbon fractions, hydroxyl content, crystallinity of carbohydrates, and porosity, in addition to simple yield of monomer sugars from polymeric holocellulose. The influence of pretreatment types, as alkaline, acid, or oxidative, is presented, using a time series of saccharification to modulate yields across a spectrum. This allows comparison at equal yields as well as among residues created upon complete saccharification. Additionally, a treatment severity index was used to equalize pretreatment parameters, although it was clear that this severity index was oversimplified using our approach, specifically for oxidative pretreatments. An additional component, laccase-catalyzed activation of the lignin binder, has also been included and its progress will be presented. The basic characterization in our work not only forms the basis for our ongoing work making solid materials from this residue matrix, but also provides valuable fundamental information on the chemical properties of saccharification residues to be used toward any co-product strategy.

(407d) Evaluation of Alternative Non-Wood Biofiber As Part of An Overall Integrated Biorefinery Approach

Huang, H.,
Tschirner, U., University of Minnesota
Ramaswamy, S., University of Minnesota
Originally presented on: 11/6/2013 9:45:00 - 10:10:00

There has been an increasing interest in the use of woody and non-woody biomass for cellulosic biofuels, bioenergy and various other bioproducts. Agricultural residues constitute a major portion, up to 2500 million dry bone tones, of the non-food biomass that is available throughout the world and it has not been effectively utilized. In addition to considering their use for cellulosic biofuels and drop-in biofuels, one of the other considerations, an equally important value-added product, is the use of cellulose fiber as an alternate source of fiber. While the total pulp and paper consumption in the world is growing, driven mostly by the growth in the developing part of the world, but the availability of good quality fiber is scarce and in high demand. For example, most of the high quality recycled fiber is increasingly being exported to China and India driving up the cost of the fiber in the U.S. This increases the need for low cost, alternative biofiber supplanting the additional need in the developing world.

In order to address the above global need for alternative biofiber, here we discuss possible process approaches that can be integrated within a biorefinery framework [1-2]. First, the current biomass resources in U.S. are summarized and the evaluation of different biomass including bagasse, reed, corn stalk, wheat straw, and rice straw etc. is carried out. Our earlier work has shown that a stand-alone green-field mill making market pulp and/or paper from agricultural residues using conventional Kraft pulping may not be feasible due to economies of scale and environmental treatment considerations [3-4]. Unlike wood biomass, collection and transportation costs of nonwood biomass for the large-scale process are much more expensive, likely leading to high overall production cost.

Here we consider alternative processes that may be more suitable at smaller scales and can be integrated with an existing biorefinery such as a corn based ethanol facility or a pulp and paper mill. Potential pulping processes for converting non-wood biomass into biofiber are explored and compared in terms of their process operating conditions, the corresponding fiber properties as well as challenges associated with chemical recovery and environmental treatment. It is to be noted that the quality of the fiber is invariably linked to the quality of the raw material as well as the process and the range of paper products that they are suitable [3-4]. Among the many possible pulping processes, the chemi-mechanical pulping processes such as the BIVIS process, the soda processes such as the Coneox/Siloxy process, organosolv processes such as the AlCell (alcohol cellulose) process of Repap, the FreeFiber, the CIMV, and the Chempolis processes, have a higher potential for applications in small-scale pulping of alternative feedstocks for biofiber. It is also possible that some of these processes can be integrated with an existing biorefinery further improving the overall viability.

References

1. H.-J. Huang, W. Lin, S. Ramaswamy, and U. Tschirner. Process modeling of comprehensive integrated forest biorefinery - An integrated approach. *Applied Biochemistry and Biotechnology*. 154(1-3):26-37, 2009.

2. H.-J. Huang, S. Ramaswamy, W. Al-Dajani, U. Tschirner. Process modeling and analysis of pulp mill-based integrated biorefinery with hemicellulose pre-extraction for ethanol production: A comparative study. *Bioresource Technology*. 101(2): 624-631, 2010.

3. Shrinath, A., U. Tschirner., S. Ramaswamy, "Economics and feasibility of a Greenfield Cereal Straw market pulp mill" Pulp and Paper Magazine of Canada 104: 10, 34-37, October 2003

4. Tschirner, U., S. Ramaswamy, S., Goel, A. "Effect of cereal straw fiber addition on papermaking furnish", Pulp and Paper Magazine of Canada 104: 10, 26-29, October 2003

(407e) Synthesis of Lithium Aluminum Silicate From Different Sources of Silica Via Sol-Gel Method

Chen, H., Texas State University-San Marcos
Lee, K. A., Texas State University-San Marcos
Oliphant, A. J., Texas State University-San Marcos
Martin, J. C., Texas State University-San Marcos
Sun, L., Texas State University-San Marcos
Originally presented on: 11/6/2013 10:10:00 - 10:35:00

This paper studied the effect of different sources of silica on synthesizing lithium aluminum silicate (LAS) via sol-gel method. Specifically, silica from ash of none treated rice husks (rice husk ash, RHA), silica from HCl treated rice husks (Silica-RH-HCl), and fumed silica were used to prepare LAS powders under various sintering temperatures, i.e., 400, 600, 800 and 1000 °C. The X-ray diffraction (XRD) characterization clearly showed that starting from 600 °C, LAS crystals (β-spodumene) formed when fumed silica and Silica-RH-HCl were used as the silica source. On the other hand, for RHA, substantial crystalline structures formed only at 800 °C or higher. The apparent reactivity of RHA is the lowest among the three sources, while the reactivity of the low cost biogenic Silica-RH-HCl was comparable to the commercial fumed silica. This apparent reactivity of silica can be mainly attributed to its crystallinity. In brief, the lower crystallinity of the silica, the higher reactivity. Scanning electron microscopy (SEM) characterization showed that regular crystals of LAS (β-spodumene) formed from all three sources of silica at 1000 °C.

(407f) Value Added Co-Product Lactic Acid Development From Lignocellulosic Biomass

Li, M., Auburn University
Tu, M., Auburn University
Originally presented on: 11/6/2013 10:35:00 - 11:00:00

Producing biofuels from lignocellulosic biomass is not economically viable despite significant process has been made in biomass pretreatment, enzymatic hydrolysis and microbial fermentation. Liquid fuel, especially ethanol, is a low value commodity product. As a result, using woody biomass to produce biofuels cost-effectively poses a great economic challenge.

Identifying value added co-products along with the production of biofuels is essential to the success of a viable biomass conversion process. Lignocellulosic biomass is primarily composed of cellulose, hemicelluloses and lignin. The technical feasibility of producing biofuels (e.g. ethanol, butanol, hydrocarbon fuels) from cellulose, has been well demonstrated. However, converting biomass hemicelluloses to fuel in high yield remains a much greater challenge.

The objective of our research is to investigate potential and promising routes for converting biomass hemicelluloses to value added co-products such as lactic acid and acrylic acid. In this work, we studied the base-catalyzed reaction of glucose, mannose and xylose to lactic acid under various catalysts. In addition, the reaction kinetics were determined and compared.

(411a) Comparative Biochemical Analysis of Biological Pretreatment of Wood Biomass With White Rot Fungi Using Sum Frequency Generation (SFG) Spectroscopy and Secondary Ion Mass Spectrometry (SIMS)

Rajangam, A., The Pennsylvania State University
Kafle, K., Pennsylvania State University
Jackson, L., The Pennsylvania State University
Maher, T., The Pennsylvania State University
Dazen, K., The Pennsylvania State University
Lee, C., Pennsylvania State University
Winograd, N., The Pennsylvania State University
Kim, S. H., The Pennsylvania State University
Curtis, W., The Pennsylvania State University
Originally presented on: 11/6/2013 8:30:00 - 8:55:00

Advanced analytical methods of Sum Frequency Generation (SFG) spectroscopy, Secondary Ion Mass Spectrometry (SIMS), InfraRed microspectroscopy are used to quantify difference in the dynamics of degradation of lignocellulosic substrates by two different wood-degrading fungi. Wood is a composite that undergoes natural degradation in different ways depending on the genetics of the degrading microorganism. This provides a means to further refine these methods of characterization while learning more about the underlying biological process. We are particularly interested in the degradation process because of its particular utility for pretreatment of biomass for the production of biofuels, but this approach to study provides a validation of the logic of depolymerization of lignocellulosic biomass that has its basis in the genetics of the organisms under study.

Wood is composite of cellulose and lignin and the lignin is the toughest biopolymer to degrade and chemical methods to remove lignin has been very difficult and often incomplete. *Phanerochaete chrysosporium* and *Ceriporiopsis subvermispora* are well characterized white rot fungi known to systematically degrade lignocellulosic biomass leaving some or most of the cellulose behind; of which *P. chrysosporium* has been used in biopulping in paper and pulp industries. Delignification is the initial process, as the recalcitrant lignin is found on the surface, coating the cellulose and hemicelluloses, through complex sequence of biochemical reactions. Both the organisms are known as efficient delignifier with varying mechanistic cellulolytic activity. *P. chrysosporium* simultaneously degrade cellulose, hemicellulose and lignin whereas *C. subvermispora* preferentially remove lignin. Number of studies have been performed to understand the expression profile - at both transcript and enzymatic levels - of lignocellulosic and carbohydrate active enzymes (CAZymes) secreted during degradation of lignocellulose by *P. chrysosporium* and *C. subvermispora*. The comparative genomic and proteomic analyses of the expression profile of *P. chrysosporium* and *C. subvermispora* suggest a differentially ordered mechanism of degradation by both the organisms reflecting the diversity and presence of some CAZymes such as CBH1/GH7, EG/GH5 and EG/GH 12 families in both these organisms.

A comparison of two fungi is undertaken in this work based on the genomic comparison where the hypothesis is that *C. subvermispora* will be more selective for lignin consumption, and *P. chrysosporium* will leave more cellulose behind. The wood chips of all the four biomass studied - Poplar, Oak, Pine and Switch grass - show differential degradation pattern with respect to the change in the orderliness of the cellulose crystals detected using SFG analysis. The orderliness could potentially reflect exposure of the underlying crystalline cellulose after delignification; the extend of lignin removed simultaneously studied by analysing the depleting lignin content through the depth of the wood sample using SIMS analysis. Preferential lignin degradation and differential pattern of degradation by each organism changes the cellulose:lignin ratio which is studied using IR-microspectroscopy. The study is perfored using 1cmx 1cm square of wood chip of 5 mm thickness from all the woody biomass and a unit leaf blade of grassy member, placed on freshly grown fungal mycelium on a culture plate. The samples were removed and sectioned into equal portion for each of the anlaysis that are collected every week for four weeks. The SFG signals are generated from; (1) the CH2 stretch from the exocyclic hydroxymethylene group (6CH2OH), and (2) the 3OH and 2OH stretch vibrations from intra-chain hydrogen bonding along the cellulose chain. In lignocellulose, the SFG signal is only generated from crystalline cellulose; there is no signal contribution from any amorphous wall polymers or absorbed water. The change in CH:OH signal reflects the long range ordering in crystalline cellulose. The data from SFG analysis of wood degradation suggest that the CH2 and OH signal ratio is consistently changing in all the wood samples while degraded by both the fungi. The hetetrogenity of differential enzymative degradation of both the fungi and the way the fibers are lignified in hardwood, softwood and grasses marks the difference in SFG signal.

SIMS analysis could chemically image the presence of the fungi and the chemical changes which occur with distribution and depth as a function of time. This may permit a chemical depth profile of the chemistry as the intrusion of fungi progresses. The SIMS analysis is used to trace the chemical signatures of lignin, cellulose, fungi, and other chemicals along the gradient. 2D and 3D Chemical imaging is being conducted with of a physically cut cross section of the sample and after etching by the primary ion beam, C60+, respectively. This is because an advanced degradation (including approx. 70% delignification) is anticipated to have gone several millimeters deep. The change in ratio of lignin:cellulose while either one is preferentially consumed in each case and during the time course of the degradation is measured using IR microspectroscopy. We anticipate a change in lignin:cellulose ratio during the course of degradation. The lignin is found only in the secondary cell walls and therefore confocal imaging is performed to image the change in the cell wall morphology and thickness in each of the condition. The confocal imaging is peformed to image the autoflourescent lignin with an excitation of laser line 514 nm, emission filter BP 530 to 600 nm. The images are being analysed to understand the physiology of delignification and resultant change in cell wall thickness and morphology.

In light of the importance of the biomass pretreatment of lignocellulose for various industrial application including making of composite from cellulosic material; establishing SFG, SIMS and IR microspectroscopy as a tool to quantify constituents of the material should find application beyond depolymerization for biofuels.

(411b) Upgrading and Pretreatment of Douglas-Fir Forest Residue for Fermentable Sugar Production

Zhu, J., USDA Forest Products Laboratory
Leu, S. Y.,
Gleisner, R., US Forest Service
Sessions, J., Oregon State University
Marrs, G.,
Originally presented on: 11/6/2013 8:55:00 - 9:20:00

Forest harvest residues can be a cost-effective feedstock for a biorefinery, but the high lignin content of forest residues is a major barrier for enzymatic sugar production. Physical fractionation was applied to a Douglas-fir (*Pseudotsuga menziesii* (Mirb) Franco var. *menziesii*) forest residue and found to be very effectiev to selectively reduce bark and ash content. The accepted residue is was used for pretreatment by sulfite pretreatment to overcome strong recalcitrance of lignocelluloses (SPORL) in a range of sulfite and acid loadings at 165 °C for 75 minutes with liquor to wood ratio of 3:1. Sodium bisulfite and sulfuric acid charge as mass fraction of oven dry biomass of 12 % and 2.21 %, respectively, was optimal in terms of enzymatic cellulose saccharification, sugar yield and formation of Hydroxymethylfurfural (HMF) and Furfural. Enzymatic glucose yield of a dry biomass was 345 g kg^{-1}, or equivalent to 82.3 % of theoretical at a cellulase (CTec2) dosage of 15 filter paper unit (FPU) per gram of glucan. HMF and furfural formation were low at approximately 2.5 g L^{-1} each in the pretreatment hydrolysate. Delignification was important to achieve good cellulose saccharification efficiency, however, and approximately 90 % hemicellulose removal is also required. A combined hydrolysis factor (*CHF*) was found to correlate well with the enzymatic hydrolysis glucose yield (*EHGY*), suggesting that *CHF* can be used to predict saccharification of forest residue for scale-up studies to reduce numbers of experiments.

(411c) Investigation On the Utilization of Grassy Biomass Grown in Heavy Metal Contaminated Soils for the Production of Sugars and Fast Pyrolytic Oil Via a Hybrid Biochemical and Thermochemical Approach

Ruiz-Felix, M. N., Villanova University
Kelly, W. J., Villanova University
Balsamo, R., Villanova University
Satrio, J. A., Villanova University
Originally presented on: 11/6/2013 9:20:00 - 9:45:00

Our society is still highly reliant on fossil fuels which are not renewable and causing environmental issues, primarily due to greenhouse gas emissions. To reduce this dependency, it is necessary to find resources that are renewable. Biomass is the only renewable source of carbon that, when utilized efficiently and optimally, can significantly help to reduce our reliance on fossil fuels. Lignocellulosic biomass is abundant, inexpensive and does not compete with the production of food crops. Nevertheless, the complex nature of biomass and many ill-defined issues related to biomass utilization pose a substantial challenge to large-scale biomass utilization. Growing biomass for energy crops, such as switchgrass, means that more land needs to be utilized. It is desired that biomass for energy crops would be grown in land areas that are considered marginal and not currently being utilized for anything productive.

At Villanova University, we are working on the development of a sustainable processing system for producing biofuels from grassy plants that can be grown on heavy-metal contaminated soils in underutilized areas such as superfund sites. Certain types of perennial grasses, such as switchgrass, have great potential to be used as bio-renewable feedstock for the production of bioenergy and biofuels due to their relatively high yields of biomass per acreage of land. Also, they have the capability of extracting heavy metals from the contaminated soil, which in turn can improve the quality of the soil. Our preliminary studies reveal great potentials of both switchgrass and timothy grass grown as soil remediating agents.

The purpose of this study was to find the optimal conditions for biofuel production from switchgrass that has been grown in soil contaminated with lead through a hybrid biochemical (i.e. enzymatic hydrolysis) and thermochemical (i.e.fast pyrolysis) approach. For the biochemical approach we are investigating the performance of enzymatic hydrolysis by using various combinations of three different enzymes, *Trichoderma Reesei*, *Aspergillus Niger* and *Humicola Grisea* for producing sugars from switchgrass that was grown in lead contaminated soils. With the thermochemical approach, we are evaluating the production of fast pyrolytic oil from the fast pyrolysis of switchgrass and the solid remaining from enzymatic hydrolysis of switchgrass. Also, the study evaluates the effects that acid hydrolysis pretreatment has on the performance of enzymatic hydrolysis and the fast pyrolysis of the biomass in producing sugars and fast pyrolytic oil, respectively.

To be reported are results of the performances of several different process configurations of combined mild-acid hydrolysis pretreatment, enzymatic hydrolysis, and fast pyrolysis for producing sugars and fast pyrolytic oil from switchgrass materials that were grown in lead-contaminated soils and un-contaminated soils. Process variables reported in this study include variation of acid concentration levels in the acid hydrolysis steps, variation of the enzymes used in enzymatic hydrolysis for the production of sugars and the variation of process temperatures of the fast pyrolysis step for the production of fast pyrolytic oil. Important results to be reported include the yields and properties of products produced from each step of the process. Properties of the solid products evaluated include results from proximate analysis (volatiles, fixed carbon and ash composition), fiber analysis (cellulose, hemicellulose, and lignin content), heating values, and elemental analysis (CHONS). On fast pyrolysis step, analysis includes the chemical product distribution of the fast pyrolytic oil. Finally, understanding the distributions of the inorganic minerals, including the heavy metal, based on various combinations of the process system configurations is another important part of the study to be reported.

(411d) Study of Lignocellulosic Biomass Pretreatment and Its Effect On Product Distribution Using Fast Pyrolysis

Garrido, R., Villanova University
Reckamp, J., Villanova University
Peterson, L., Villanova University
Ruiz-Felix, M. N., Villanova University
Satrio, J. A., Villanova University
Originally presented on: 11/6/2013 9:45:00 - 10:10:00

Study of Lignocellulosic Biomass pretreatment and its effect on product distribution using Fast Pyrolysis

Rene Garrido, Joseph Reckamp, Lindsay Peterson, Maria Nydia Ruiz-Felix, and Justinus A. Satrio

Department of Chemical Engineering, Villanova University

800 East Lancaster Avenue, Villanova, PA 19041

ren.garrido@villanova.edu; mruizfel@villanova.edu; justinus.satrio@villanova.edu

Energy based in biomass is a source of renewable energy from organic and inorganic matter formed in a biological or mechanical process. Although fossil fuels, such as petroleum, oil and coal, have their foundation in ancient biomass, they have been "out" of the carbon cycle for a long time. Their combustion consequently disturbs the carbon dioxide content in the atmosphere.

Lignocellulosic biomass can be classified as natural (without human intervention), residual (by-product of waste generated by agricultural activities, forestry, and industry wood processing), and energy-based crops (for the production of biofuels); it is the only carbon-containing renewable energy source.

The complex structure of lignocellulosic biomass has been extensively studied and poses highly valuable carbohydrates (cellulose and hemicelluloses) and lignin, whereas there is a substantial challenge to the large-scale biomass utilization. The carbohydrate polymers are strongly bound to the lignin, which makes the separation process technically difficult.

While it is scientifically viable to exploit lignocellulosic materials and organic wastes into energy, chemicals and fuels, it is far from an economic process and the cost needs to be lowered. It also must be demonstrated that a commercial scale biomass utilization process is environmentally sound, and there is a need to find an inexpensive and widely available lignocellulosic source of biomass

This study investigates common Reed Grass *(Phragmites Australis)*, Paper Mill Sludge (PMS), and Pinewood as an example of different sources of lignocellulosic biomass. It was found that, even though they all have a high probability for bio-oil conversion its high moisture content and, in some cases, a high ash content in feedstock results in a low yield and quality of the bio-oil. The removal of these recalcitrant is a key element to make a more desirable feedstock for biomass conversion. This investigation has revealed that using torrefaction and hydrolysis as pretreatments before pyrolysis shows an interesting switch in the kinetics reaction of all feedstocks producing predominantly levoglucosenone as one of the main products of pyrolysis.

Analytical pyrolysis combined with gas chromatography and mass spectrometry (Py-GCMS) is used to study the chemical composition of volatile products from previously hydrolyzed and torrefied biomass. Effects of the severity of hydrolysis (0 – 20% H_3PO_4 solutions) pre-treatment and torrefaction temperatures on the chemical composition and yield distribution of the fast pyrolysis products were evaluated and it was possible to determine the best pretreatment condition combinations for each feedstock. The chemical composition study of pyrolyzed biomass would then aid in the understanding of the chemistry behind the different compounds to further enhance the process of upgrading bio-crude oil. By this, ground work is laid as a preliminary attempt to understand how different feedstocks can be used for biomass conversion.

(411e) Composition and Structure Analysis of Sweet Sorghum Stalk

Li, J., Tsinghua University
Yan, Z., Tsinghua.University
Li, S. Z., Institute of Nuclear and New Energy Technology, Tsinghua University
Originally presented on: 11/6/2013 10:10:00 - 10:35:00

Facing to the rapid depletion of fossil resource, a renewable and sustainable resource of energy is expected. Sweet sorghum as a multi-plantform energy crop is regard as a promising feedstock for producing biofuels. The recalcitrant characters of different parts of sweet sorghum stalks are major factors to affect the bio-degradation performance of the feedstock. In this study, the chemical compositions of the sweet sorghum stalks including leaf, bark, pith, kont, and different height parts of stalks were analyzed. Furthermore, their chemical and physical structures were characterized by IR, XRD and SEM. The results showed there are not only significant differences in the lignin content and S/G ratio of lignin for different parts of stalks, but also significant difference in crystallinity of celluloses. The further enzymatic hydrolysis experiment demonstrated the cellulose conversion depended on the lignin content.

(411f) Optimization of Dilute Acid and Liquid Hot Water Pretreatment of Plantain Rachis for Monomeric Sugars Production

Ruiz, A., Universidad Nacional de Colombia
Torres, A., Universidad Nacional de Colombia - Sede Medellin
Originally presented on: 11/6/2013 10:35:00 - 11:00:00

Optimization of Dilute Acid and Liquid Hot Water pretreatment of Plantain Rachis for Monomeric Sugars Production

Torres-Conde, A. , Universidad Nacional de Colombia – Sede Medellín. Facultad de minas

Ruiz-Colorado, A. , Universidad Nacional de Colombia – Sede Medellín. Facultad de minas

Garzón, L. , Instituto Colombiano del Petróleo (ICP)

Keywords: Monomeric sugars, hydrolysis, pretreatment, plantain.

Plantain rachis is an agricultural waste that represents an abundant cellulosic material and it has potential as a feedstock for producing cellulosic biofuels. Efficient conversion of lignocellulosic biomass into monomeric sugars requires particle size reduction, pretreatment, and enzymatic hydrolysis. Liquid hot water (LHW) and dilute acid (DA) pretreatments allow a selective solubility of hemicellulose that improves the accessibility of enzymes that hydrolyze cellulose. We present the optimization conditions of LHW pretreatment for temperatures between 170 and 200 °C, at times between 15 and 25 min and solids loading of 7 and 17% dry matter (w/w), optimal conditions of DA pretreatment are temperatures between 160 and 200 °C, at times of 5 and 10 min and H_2SO_4concentrations of 1 and 2% (w/w). The chosen design was a response surface - central composite design unblocked with 20 experiments including 6 repetitions at central point conditions for both pretreatments.

Chemical composition of pretreated and unpretreated plantain rachis was obtained following National Renewable Energy (NREL) Laboratory Analytical Procedures (LAPs). Unpretreated biomass was milled to pass through 20 mesh, moisture and ash content were measured, and a two-step (water-ethanol) extraction was done to remove soluble nonstructural components. Pretreated samples were subjected to the same analysis without taking into account the extraction step.

Saccharification (SAC) reactions was held for 72 hours and stirred (150 rpm) in 250 mL flasks at a pretreated unwashed solid loading of 10% dry matter(w/w), 0.05 M Na-citrate (pH 4.8) and cellulase dose of 66.07 FPU/g cellulose to bring the reactive volume to 100ml.

Regarding LHW, the highest saccharification yields of 0.2807 g glucose/g initial biomass were obtained at 200 °C, 40 min and a solid loading of 16% dry matter (w/w), while the best condition through using DA for yield 0.2532 g glucose/g initial biomass was 200 °C, 8 min and H_2SO_4concentrations of 2.34% (w/w).

References.

[1] M. Zeng and N. Mosier, "Microscopic examination of changes of plant cell structure in corn stover due to hot water pretreatment and enzymatic hydrolysis," *Biotechnology and bioengineering*, vol. 97, no. 2, pp. 265–278, 2007.

[2] A. Kunamneni and S. Singh, "Response surface optimization of enzymatic hydrolysis of maize starch for higher glucose production," *Biochemical Engineering Journal*, vol. 27, no. 2, pp. 179–190, Dec. 2005.

[3] F. Zimbardi, E. Viola, F. Nanna, E. Larocca, M. Cardinale, and D. Barisano, "Acid impregnation and steam explosion of corn stover in batch processes," *Industrial Crops and Products*, vol. 26, no. 2, pp. 195–206, Aug. 2007.

[4] L. Oliveira, N. Cordeiro, and D. Evtuguin, "Chemical composition of different morphological parts from 'Dwarf Cavendish'banana plant and their potential as a non-wood renewable source of natural products," *Industrial Crops and Products*, vol. 26, pp. 163–172, 2007.

[5] Y. Kim, N. S. Mosier, M. R. Ladisch, V. R. Pallapolu, Y. Y. Lee, R. Garlock, V. Balan, B. E. Dale, B. S. Donohoe, T. B. Vinzant, R. T. Elander, M. Falls, R. Sierra, M. T. Holtzapple, J. Shi, M. a Ebrik, T. Redmond, B. Yang, C. E. Wyman, and R. E. Warner, "Comparative study on enzymatic digestibility of switchgrass varieties and harvests processed by leading pretreatment technologies.," *Bioresource technology*, vol. 102, no. 24, pp. 11089–96, Dec. 2011.

[6] M. Santa-Maria, A. A. Ruiz-Colorado, G. Cruz, and T. Jeoh, "Assessing the Feasibility of Biofuel Production from Lignocellulosic Banana Waste in Rural Agricultural Communities in Peru and Colombia," *Bioenergy Research*, Apr. 2013.

(468a) Diversifying Renewable Feedstocks for Biobased Polymers and Applications

Zhang, J., college of Engineering and Architecture, Washington State University

Originally presented on: 11/6/2013 12:30:00 - 12:55:00

Compared with petrochemical polymers, biobased polymers are very scarce in terms of variety and production volume. Only a few biobased polymers are available with sizable scales of commercial production, and they only find some niche applications due to limited performance and relatively high costs. In order for biobased polymers to replace conventional polymers in broad applications, developing new polymers with diverse properties is necessary. In this work, we made a series of attempts to explore the use of several renewable feedstocks in developing new polymer applications. The renewable feedstocks used include rosin, plan oils, lignin, terpenes, etc. Syntheses of epoxies, curing agents, poylurethanes and block copolymers using these renewable feedstocks were studied. By manipulating the molecular structures of monomers and polymers, the resulting polymers exhibited a broad spectrum of properties.

(468b) Development of Biochar As a Promising Carbon-Based Candidate in Electric Double Layer Applications

Dehkhoda, A. M., The University of British Columbia
Ellis, N., The University of British Columbia
Gyenge, E., The University of British Columbia
Originally presented on: 11/6/2013 12:55:00 - 13:20:00

Biochar, a residual by-product of fast pyrolysis of woody biomass, is studied as a low-cost, novel, and renewable carbon-based electrode material for electric double layer (EDL) based applications. The main EDL-based applications include water and wastewater purification (e.g., electrosorption of industrial effluents, capacitive deionization of brackish water) and energy storage capacitor fabrication (e.g., supercapacitor fabrication). The activation of biochar for increasing the surface area, tuning the pore structure, and modifying the surface chemical groups is investigated for developing biochar as a promising carbon-based candidate for EDL-based applications. Based on the combination of chemical (i.e., impregnation in 7 M KOH) and thermal activation (i.e., carbonization at 675 or 1000^{0}C under N_2 atmosphere), the surface area and the total pore volume of the untreated biochar increased from 1.66 m^2 g^{-1} and virtually negligible up to 990 m^2 g^{-1} and 0.90 cm^3 g^{-1}, respectively. Different activation conditions resulted in activated biochar samples with various properties such as structure, surface area, porosity, and functional groups. These properties have been studied through FT-IR, XRD, and Elemental Analyses. To test the EDL performance of activated biochar samples, electrodes were fabricated by spraying the activated biochar on a flexible thin Nickel mesh (150 x 150 mesh size) as current collector. To increase the adhesiveness of activated biochar on Nickel mesh, different concentrations of Nafion® (a cation exchange binder) from 0 to 30 wt.% have been applied and the subsequent effect on the EDL performance was investigated. This electrode fabrication technique can potentially overcome major operational difficulties in commercial EDL processes units such as significant potential and pressure drop of the electrolyte flow through electrode. Furthermore, This electrode fabrication technique could be advantageous for the design and assembly of capacitors with various configurations, e.g., stacked or rolled screens.

The electrochemical properties of the activated biochar electrodes were investigated through Cyclic Voltammetry (CV) and Galvanostatic charge/discharge investigations. For comparison, electrodes have also been made from untreated biochar (as-received biochar) and a commonly used carbon black candidate in EDL applications, i.e., Vulcan XC-72. Electrodes prepared from untreated biochar showed negligible EDL behaviour as studied by CV analyses. However, the electrodes prepared from the activated biochar samples showed typical EDL behaviour as in form of a rectangular-based CV diagram. These results confirm the importance of chemical activation on the EDL performance of biochar electrodes. CV analysis was conducted at sweep rates in the range of 1-50 mV/s to quantify the total capacitance of each electrode. The electrolyte was a mixture of 0.1 mol L^{-1} NaCl in 0.1 mol L^{-1} NaOH. The maximum capacitance obtained was 167 $F.g^{-1}$ for the electrode prepared from activated biochar at 675^0C and 2 hour residence time under 258 $mL.min^{-1}$ N_2 flow. Results have further suggested that increasing the Nafion® content during electrode preparation yields lower total capacitances, i.e., 43 $F.g^{-1}$ capacitance reduction while increasing the Nafion® content from 0 to 30 wt.%. The activated biochar electrodes have also showed promising charge/discharge behaviour through Galvanostatic analyses. The maximum discharge capacitance of activated biochar electrodes was 55.4 $F.g^{-1}$ which is considerably higher than that of Vulcan XC-72 electrode prepared at identical conditions. Furthermore, the total capacitances of the activated biochar electrodes are competitive with much more expensive systems such as carbon nanotubes CNTs and graphene-based electrodes in the literature [1-3].

The activated biochar electrodes are novel, green, and low-cost candidates for EDL-based applications with promising characteristics such as high surface area, high content of carbon oxygen groups, specific type of structure, and comparable electrical conductivity with conventional carbon-based electrodes. Developing this new generation of electrodes can further improve the economical feasibility of commercial units of EDL-based processes, e.g., electrosorption units and supercapacitor fabrication, due to much less costly preparation methods compared to those of promising carbon electrodes (e.g., carbon aerogels and carbon nano-tubes) in the literature. Moreover, the value-added utilization of biochar as carbon-based material in EDL applications would increase the overall economic outlook of woody biomass pyrolysis.

References

[1] F. Alvi, M. K. Ram, P. A. Basnayaka, E. Stefanakos, Y. Goswami, and A. Kumar, "Graphene–polyethylenedioxythiophene conducting polymer nanocomposite based supercapacitor," *Electrochimica Acta*, vol. 56, no. 25, pp. 9406–9412, Oct. 2011.

[2] Y. Honda, T. Haramoto, M. Takeshige, H. Shiozaki, T. Kitamura, and M. Ishikawa, "Aligned MWCNT Sheet Electrodes Prepared by Transfer Methodology Providing High-Power Capacitor Performance," *Electrochem. Solid-State Lett.*, vol. 10, no. 4, pp. A106–A110, Apr. 2007.

[3] Y.-K. Hsu, Y.-C. Chen, Y.-G. Lin, L.-C. Chen, and K.-H. Chen, "High-cell-voltage supercapacitor of carbon nanotube/carbon cloth operating in neutral aqueous solution," *J. Mater. Chem.*, vol. 22, no. 8, p. 3383, 2012.

(468c) Biochar in Biocomposite Uses

Misra, M., University of Guelph
Mohanty, A. K., University of Guelph
Anstey, A., University of Guelph
Mohanty, A. K., University of Guelph
Misra, M., University of Guelph
Vivekanandhan, S., University of Guelph
Originally presented on: 11/6/2013 13:20:00 - 13:45:00

Biochar is a charcoal-like product of the pyrolysis of biomass, and is emerging as an important bio-based resource. Biochar is very rich in carbon, and has been used as a soil amendment due to its ability to increase soil fertility and water retention. Biochar and its pyrolysis co-products, bio-oil and syngas, are also being studied for their application in the biofuel industry. This study investigates the potential of biochar as a structural biomaterial; specifically as a filler material in nylon polymer biocomposites. Composites were prepared from nylon-6 polymer and wood-based biochar using a twin-screw micro compounder and injection moulding. The biochar was used in its raw state (as-received from the manufacturer) and no compatibilizers were used in the composite. A range of processing and moulding temperatures, durations and pressures were used to optimize the quality of the processed composite material. The produced material was characterized through several different methods. Tensile, flexural and impact testing was used to analyze its mechanical properties. The melt flow index was analyzed for various loadings of biochar. The thermal behaviour of the material was also studied, including the coefficient of thermal expansion as well as the thermal conductivity. Scanning electron microscopy (SEM) was used to analyze the microstructure characteristics of the composite fracture surfaces. The study found that the unmodified biochar was not very compatible with the nylon-6, as the analyses showed that there was poor interfacial adhesion between the polymer matrix and the biochar particles. The biochar was also only able to be fully incorporated by the nylon matrix in relatively low loadings (under 15%). Future studies will investigate the modification of the biochar material with the goal of increasing interfacial adhesion and interaction with the nylon matrix.

Acknowledgements: The authors are thankful to the Ontario Ministry of Agriculture, Food and Rural Affairs (OMAFRA)/University of Guelph Research Program under Product Development and Enhancement through Value Chains Research Theme and the Ministry of Economic Development and Innovation (MEDI), Ontario Research Fund - Research Excellence Round 4 program, for their financial support to carry out this research.

(468f) Chemical Activation of Biochar Made From Oil Palm Kernel Shell and Its Application to Adsorbent for p-Nitrophenol

Uemura, Y., Universiti Teknologi PETRONAS
Afif, H., Universiti Teknologi PETRONAS
Osman, N., Universiti Teknologi PETRONAS
Riaz, N., Universiti Teknologi PETRONAS
Originally presented on: 11/6/2013 14:35:00 - 15:00:00

Lignocellulosic biomass is one of the most abundant biomass resources on the earth and can be used as a feed-stock for preparing fuels and chemicals. Some of these technologies are still under development. Due to its availability in Malaysia, oil palm waste is considered to be the best biomass waste. In 2010, Malaysia was the second largest producer of palm oil, producing 17.8 million tons, or 39% of the total world supply. Indonesia was the world's largest producer of palm oil, producing 22.2 million tonnes of oil, or 48% of the total world supply. In 2010, productive oil palm plantations in Malaysia covered 4.9 million hectares, a 3.3 % increase from 2009, when productive oil palm plantations covered 4.7 million hectares. Types of biomass produced by the oil palm industry include empty fruit bunches (EFB), mesocarp fiber, kernel shells, fronds and trunks. EFB, mesocarp fiber and kernel shells are used or discarded at palm oil mills, while fronds and trunks are used or discarded at the plantations. The total amount of the biomass residues from the oil palm industry amounts 18.2 Mtoe (million ton oil equivalent). Since the current primary energy supply in Malaysia is roughly 70 Mtoe (million tons of oil equivalent), the total oil palm biomass energy potential of 18.2 Mtoe may contribute considerably to decreasing the consumption of fossil fuels (natural gas, coal and oil). As one of the options to utilize lignocellulosic biomass, the authors have carried out pyroslysis oil production from oil palm kernel shell (PKS), and have published some papers (Proceedings for the 19th European Biomass Conference, VP2.5.33 (2011); *Journal of the Japan Institute of Energy*, **91**, 954-959 (2012)). Pyrolysis of PKS produces not only pyrolysis oil but also biochar at the yield of 30-40 %. In this study, the biochar, which was prepared by microwave pyrolysis of PKS, was chemically activated, and utilized as an adsorbent of p-nitrophenol dissolved in water. 10 g of the biochar was impregnated with $ZnCl_2$ aqueous solution at 298 K for 1 h, then dried in an oven at 378 K overnight. The dried-up sample was heated up to the desired temperature (673 or 773 K) at 3 deg/min, and was maintained at the temperature for 1 or 1.5 h. The mass ratio of biochar to $ZnCl_2$ was 10:1 and 10:10. The eight different activated biochars prepared were used for adsorption of p-nitophenol from aqueous solution. The biochars were also characterized by BET, FTIR, XRD, SEM and TEM. The biochar activated at 673 K for 1 h with the biochar to $ZnCl_2$ mass ratio of 10 to 1 showed the best adsorption capability among the eight samples prepared. The adsorption isotherm was the Freundlich type.

(521a) Characterization of Bio-Oil, Bio-Char from Biomass Fast Pyrolysis and Bio-Char Application into Soil

Wang, Y.,
Liu, R.,
Yin, R.,
Wu, L.,
Originally presented on: 11/6/2013 12:30:00 - 12:42:00

Fast pyrolysis of biomass was carried out in a fluidized bed reactor. Three kinds of products including bio-oil, biochar and non-condensible gases were obtained. Characterization of bio-oil, bio-char from biomass fast pyrolysis and bio-char application into soil were investigated. Bio-oil is a clean, cost effective liquid alternative to replace fossil fuel for 'green power' generation, transportation and district heating. However, the properties and chemical composition of bio-oil are distinct from petroleum fuel. In general, due to its relatively high water and oxygen content, bio-oil is acidic and not stable when heating. Components of bio-oil from fast pyrolysis of rice hull and their relative mass contents were detected by GC–MS method. It is shown that fast pyrolysis is a complex process. 112 types of chemical compounds were detected. The ash and fixed carbon content increased and volatile component content decreased for all the three feedstock including elm wood, pine wood and rice hull after pyrolysis. The rice hull biochar had the highest ash content and the lowest fixed carbon and volatile component, which was 50.94%, 33.35% and 13.11%, respectively. The pH value of biochar was from neutral to alkalinity. Of the three kinds of biochar, the pH of pin wood biochar(9.06) was higher than that of rice hull biochar(7.49) and elm wood biochar(6.35). Therefore, rice hull biochar and pine wood biochar could be applied to acid soil improvement. Rice hull biochar had the highest specific surface area (BET) (78.15 m3/g) and apparent densities (418.3 kg/m3), and SEM analysis also confirmed that rice hull biochar had the most micropores. Fourier transform-infrared (FT-IR) demonstrated that the functional groups of three kinds of feedstock were similar, and also similar of three biochar except that rice hull biochar had extra arene functional groups. Results of two-month soil experiment showed that rice hull biochar(4%) could increase soil nutrition including N, K, Mg and Mn, especially for K content of soil which increased from 0.055 to 0.417mg/kg. The soil carbon increased 1.69 times, and the pH value increased from 3.33 to 3.62, but there was no significant effect to phosphorus. The most important thing is that rice hull biochar reduced the Pb and Al availability to some extent, so it could be used to improve the soil quality that was contaminated by heavy metals. Although application biochar into soil has many advantages, a long-term field experiment and the effect of production of biochar on environment need to be estimated before application of biochar in a large scale.

(521b) Process Development and Optimization of Cellulosic Derived Sugars from Biomass for Biofuels and Chemicals Applications

Parekh, S., Sweet Water Energy

Originally presented on: 11/6/2013 12:42:00 - 12:54:00

The recent volatility of petroleum prices is stimulating the biorefining industry to diversify its raw material feedstock, using a broad-based sugar platform technology. Corn silage and energy crops such as sweet sorghum are widely recognized as the promising high-volume, low-cost lignocellulosic feedstocks from which to derive C5- and C6-rich sugar streams that can be bioconverted into biofuels and chemicals.

This presentation will review Sweetwater Energy's overall approach and concept of developing a process based on a decentralize model, whereby a modular, feedstock-flexible bioprocessing machine pretreats and processes ligocellulosic biomass, and generates adequate quantities of C5 and C6 sugar-rich streams at the biomass-generation site. The sugars are then transferred to a regional refining site for the production of biofuels. This talk will highlight: 1) the importance of biomass harvesting and storage technologies from the field to the bioprocessing site, 2) characterization of the physical and chemical properties that enable optimum bioconversion process, 3) specific pretreatment conditions and various process strategies for deriving streams rich in C6 and C5 sugars for biofuels production, and 4) review recent advances made at Sweetwater Energy in sugar platform technologies, 5) process scale up thereby accelerating process commercialization pretreating a wide range fof biomass into cellulosic sugars that can be used for various applications in ethanol chemicals and advanced fuels, such as bio-jet fuel.

(521c) Enzymatic Hydrolysis of Old Corrugated Cardboard (OCC) Fines from Recycled Linerboard Mill Waste Rejects

Min, B. C., SUNY College of Environmental Science and Forestry
Bhayani, B., Empire State Paper Research Institute
Ramarao, B. V., ESPRI, SUNY College of Environmental Science and Forestry
Originally presented on: 11/6/2013 12:54:00 - 13:06:00

During the manufacture of linerboard from recycled OCC (old corrugated cartons), a significant fraction of the fibers are broken into small fine particles. These fines are rejected into the solid waste stream placing a cost for landfilling and posing environmental problems. The fines rejects are composed primarily of cellulose which can be hydrolyzed into sugars for possible fermentation into biofuels or other products. These fines have some advantages such as low cost and without any necessity of complex pretreatment processes characteristic to reduce lignocellulosic biomass. The enzymatic hydrolysis of reject fines from a recycled OCC mill was studied for producing sugars and the methods of reducing costs were considered. Three commercially available enzyme mixtures were tested. Cellulase from Trichoderma Reesei was found to be the most effective at loading levels of 5 – 100 FPU (/g of oven dry mass). The maximum hydrolysis yield was 45% sugar (g/g-OD fines) with 50 FPU. Since the fines rejects contain fillers (upto 30% by mass), the enzyme loading was lowered by using surfactants to cover the mineral surfaces. Reductions of the enzyme dosage were observed to obtain high hydrolysis yield using non-ionic surfactants (Tween 80, CTAB) or different pH buffer. As a result, Tween 80 reduced enzyme requirement to 30 FPU and Combination Tween 80 and pH 4 sodium acetate buffer increased efficiency of low enzyme dosage.

(521d) Effects of Fuel Properties on Natural Downward Smoldering of Piled Biomass Powder: Experimental Investigation

He, F.,
Gao, Z.,
Luo, B.,
Li, Z., Shandong University of Technology
Yi, W., Shandong University of Technology
Originally presented on: 11/6/2013 13:06:00 - 13:18:00

Effects of fuel properties on natural downward smoldering of piled biomass powder: experimental investigation

Fang He, Zhenqiang Gao, Bin Luo, Zhihe Li, Weiming Yi

Shandong University of Technology, Zibo, Shandong, 255049 P.R. China

Abstract: In order to obtain data for modeling validation of one-dimensional smoldering and combustion of biomass, effects of fuel type, moisture content, and particle size on natural downward smoldering of biomass powder have been investigated experimentally. A cylindrical reactor (inner size Φ26 cm — 22 cm) was built and powders of corn stalk, pine trunk, pyrolysis char and activated char from corn stalk were prepared.

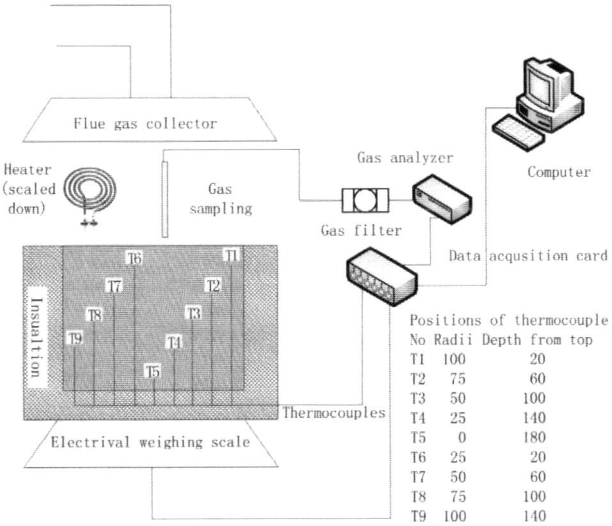

Fig. 1? Schematic of the experimental set-up for the natural downward smoldering

Smoldering characteristics of the four materials, different moisture contents (between 0% and 30%, wet basis, corn stalk powder) and different particle sizes (between 20 and 100 mesh, dry corn stalk powder) were examined. Results showed that: 1) The maximum temperature in the fuel bed is only slightly affected by fuel type and particle size. It increases gradually for original biomass and decreases slowly for chars with the development of the process. 2□Propagation velocity of the char oxidation front is dominated by carbon density and ash content. And it is almost not affected by moisture content and particle size. 3) There is always a drying front in the fuel bed during smoldering of hydrocarbons even if the original material is dry. Propagation velocity of the drying front is affected by moisture content significantly, decreases from more than 10 to 3 times of char oxidation velocity with the increase of moisture content from 0 to 20%. 4) Cracks inside the fuel bed of are dominated by fuel type and particle size.

Typical results:

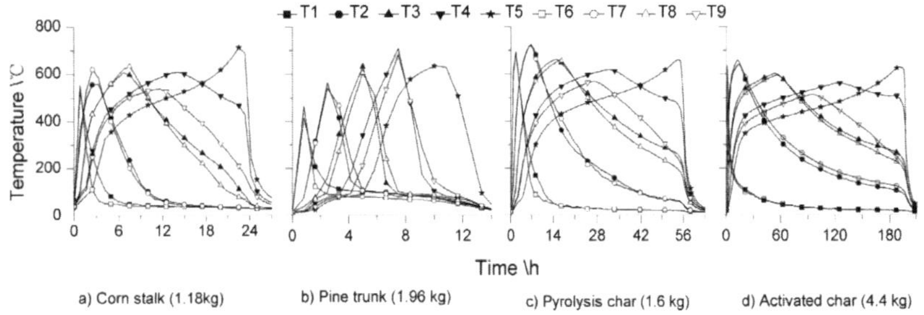

Fig. 2 ? Effects of the fuel type on the temperature of smoldering

Temperature histories of smoldering of the four fuels are presented in Fig. 2. Maximum temperatures from thermocouples are between 450 and 750 ℃ varies with their depths.

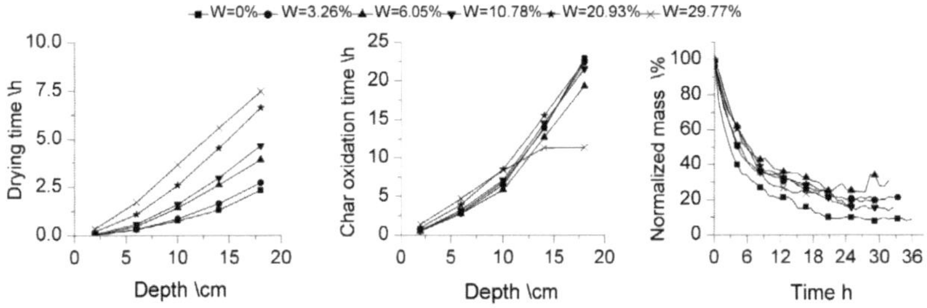

Fig. 3 ? Effects of moisture content on propagation and mass loss rate of smoldering

The time it takes for the thermocouples to reach the given temperatures□101 ℃for drying and the maximum for char oxidation) at moisture content between 0 and 30 % is presented in Fig. 3. It shows that the drying time increases significantly with the increasing moisture content. However, there is no obvious regularity between the char oxidation time and the moisture content.

Highlight:

1. Natural downward smoldering of four materials, different moisture contents, and different particle sizes were investigated.

2. Propagation velocity of the char oxidation front differs significantly from that of the drying front.

3. Carbon density and ash content of fuel dominate propagation velocity of the char oxidation front.

4. Fuel bed of finer material is apt to have bigger cracks.

Keywords: fuel type, moisture content, particle size, biomass, smoldering

(521e) Experimental Study of Bio-oil Production from Corn Stalk Using a Fluidized Bed under Hot Flue Atmosphere

Li, Z., Shandong University of Technology
Niu, W.,
Yi, W., Shandong University of Technology
Li, Y.,
Fu, P., Shandong University of Technology
Zhang, D.,
Wang, J., Shandong University of Technology
Bai, X., Shandong University of Technology
Originally presented on: 11/6/2013 13:18:00 - 13:30:00

Experimental study of Bio-oil Production from Corn Stalk using a Fluidized bed under hot flue atmosphere

Zhihe Li[a], Weisheng Niu[b], Weiming Yi[a]*, Yongjun Li[a], Peng Fu[a], Deli Zhang [a], Jing Wang[a], Xueyuan Bai[a] *

[a]*School of Agricultural and Food Engineering, Shandong University of Technology, Zibo 255049, China*

[b]*School of Engineering, Shenyang Agricultural University,Shenyang110866, China*

*Corresponding author. Tel.: +86 533 2786308; Fax: +86 533 2786158

Abstract

In order to investigate the effects of pyrolysis temperature on the product yields, Â fast pyrolysis experiments for bio-oil production have been conducted on corn stalk using a novel fluidized bed reactor under hot flue atmosphere at the selected temperature range of 450-520℃. During experiments, un-condensable gases were detected on-line with Gasboard-3100. The bio-oil obtained under the experimental conditions that gave maximum oil yield were used for further characterization.The chemical compositions of the bio-oil were determined by GC-MS. Physical analysis of the bio-oil such as pH value, viscosity and calorific value were determined with PHS-2F pH meter (Shanghai Precision and Science Instrument Co., LTD), NDJ-1 rotational viscometer (Shanghai Changyi Gealogical Instruments Co., Ltd) and WZR-1T oxygen combustion bomb (Changsha Bente Apparatus), respectively.

The results show that maximum bio-oil yield of 54.1wt% is achievedd at 500℃. The main components of un-condensable gaseous mixture includes Methane, Hydrogen, Carbon dioxide and Carbon monoxide. Bio-oil is a complex compound which contains many chemical species and it is an acidic, high viscosity, low heat value and low volatility liquid in comparison with diesel fuel. The properties change rapidly during storage condition. Its utilization as fuels requires some form of upgrading to improve storage stability and heat value and reduce the water content, viscosity and acidity.

Key words: Pyrolysis; Fluidized bed, Corn stalk; Bio-oil; Flue

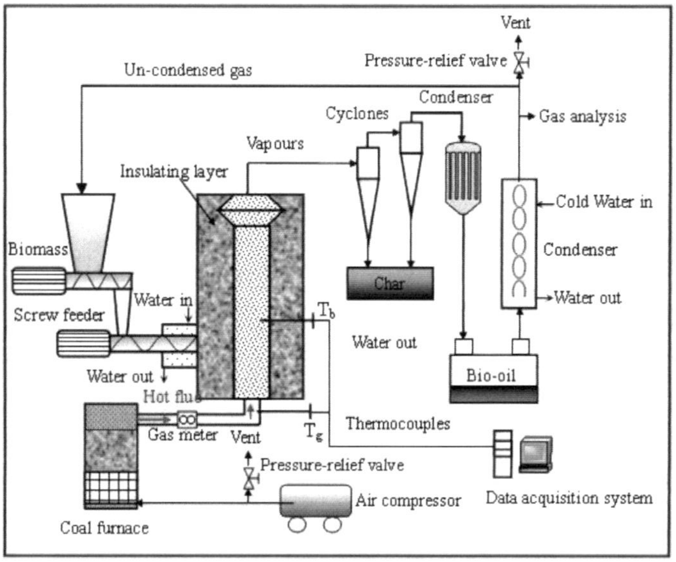

Fig. 1. Schematic diagram of the fluidized bed experimental system

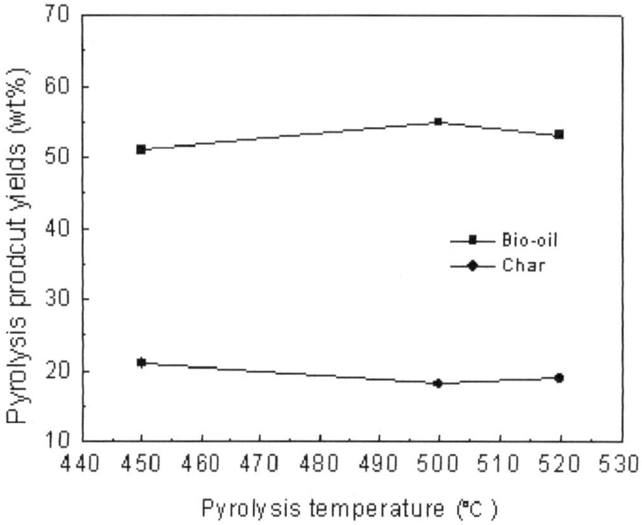

Fig. 2 Product yields of corn stalk at selected pyrolysis temperature

(521f) Quantum Chemistry Study on Adsorption of Gases from Biomass Pyrolysis on CaO Surface

Zhao, B., Energy Research Institute of Shandong Academy of Sciences
Zhang, X., Energy Research Institute of Shandong Academy of Sciences
Chen, L., Energy Research Institute of Shandong Academy of Sciences
Yi, X.,
Tian, Y.,
Chen, G.,
Originally presented on: 11/6/2013 13:30:00 - 13:42:00

Biomass pyrolysis/gasification with in-situ CO_2 absorption is a promising technology to produce the rich-hydrogen fuel gas, which can be used for wide range of applications as well as the operation of fuel cell, the generation of synthetic natural gas and hydrogen, or the heat supply through combustion. It is necessary to investigate the effects of the reaction atmosphere on CaO carbonation performance due to its role to enhance H_2 production reactions. The adsorption of CO, O_2, H_2O(g) , CO_2 , CH_4 and H_2 on CaO surface were studied by quantum chemistry method in this paper. The corresponding microscopic parameters, such as adsorption energies, adsorption distances, adsorption bond orders and changes of net charge of gases on CaO surface were calculated with the soft of material studio. The curves of binding energies between gases and CaO surface were fitted well using the mathematic function. It was indicated that the adsorption of CO_2 on CaO surface was the stronger than those of H_2O, CO, O_2, CH_4 and H_2.

(580a) Influences of Process Parameters On Cell Growth and Product Distribution of Clostridium Autoethanogenum

Xu, H., University of Chinese Academy of Sciences
Liang, C., Guangzhou Institute of Energy Conversion, Chinese Academy of Sciences
Xu, J., Guangzhou Institute of Energy Conversion, Chinese Academy of Sciences
Gao, Y., Guangzhou Institute of Energy Conversion, Chinese Academy of Sciences
Zhuang, X., Guangzhou institute of energy conversion,Chinese Academy of Sciences
Yuan, Z., Guangzhou Institute of Energy Conversion, Chinese Academy of Sciences
Originally presented on: 11/6/2013 15:15:00 - 15:40:00

Ethanol production from syngas or CO-rich waste gas fermentation is a promising technique and has received much attention due to its advantages. Up to date several acetogens have been isolated and demonstrated to have the ability of converting syngas to ethanol. Compared to other acetogens, *Clostridium autoethanogenum* is less studied, and the ethanol it produced appears to be little. In this study, influences of several process parameters on cell growth and products distribution of *Clostridium autoethanogenum* were investigated. *Clostridium autoethanogenum* was grown in serum bottles, aluminum bags and a 3L-fermentor with modified medium, respectively. Results show that pH and Oxidation-Reduction Potential(ORP) play an important role in the process of ethanol production. pH and ORP decreased rapidly accompanying with cell growth and acetic acid production, and then increased while ethanol concentration started to increase. The maximum ethanol concentration obtained in the 3L-fermentor was 2.86g/L with continuous CO supply, while in an aluminum bag it reached the level of 3.46 g/L after feeding the bag with 1L CO twice, which is 340% or 430% greater than the maximum ethanol concentration achieved in other studies. In addition, *Clostridium autoethanogenum* can grow on xylose, but produce little ethanol with it. Since hydrolysate of bagasse treated with liquid hot water contains xylose, it was used to prepare the growth medium for *Clostridium autoethanogenum*. After growing in the hydrolysate medium for 2 days, *Clostridium autoethanogenum* was transferred to an aluminum bag filled with 1L CO. 3.13g/L ethanol was detected in the medium when CO was exhausted, indicating that the medium made of bagasse hydrolysate didn't inhibit *Clostridium autoethanogenum* from growing and converting CO to ethanol. A new technique combining CO fermentation with lignocellulose hydrolysis was suggested in the end.

(580b) Design of Lignin-Based Process Plants for Production of Platform Chemicals

Palanki, D. S., University of South Alabama
Curow, T.,
Hudson, A.,
Balzaretti, D.,
Originally presented on: 11/6/2013 15:40:00 - 16:05:00

There is current interest in utilizing biomass as a sustainable, renewable alternative to fossil-based chemistry. It is estimated that in 2025, up to 30% of raw materials for the chemical industry will be produced from biomass. While significant research has been conducted on the production of ethanol fuel from starch-rich components of biomass, current research is focused on lignocellulosic biomass, which is more abundant and relatively cheap compared to starchy biomass. However, most biorefinery schemes focused on utilizing cellulose and hemicellulose to produce ethanol, leaving lignin as a waste product. Furthermore, the pulp-and-paper industries also generate significant quantities of lignin as a waste product. In this paper, it is shown how the Douglas hierarchy can be utilized to synthesize a process flow sheet that utilizes lignin as a feed material to produce several different platform chemicals that are used as starting materials for the manufacture of plastics, synthetic fibers, paints, resins, adhesives, solvents, refrigerants, pigments and dyes. Simulations are conducted in ASPEN Plus to determine the material and energy balances of these flow sheets. A mass and energy balance analysis indicates that these conceptual processes have favorable energetics and has economic potential.

(580c) Exploring 3D Nanostructure and Topochemical Distributions of Lignocellulosic Biomass and Their Impact On Biomass Pretreatment

Ramarao, B. V., ESPRI, SUNY College of Environmental Science and Forestry
Ramaswamy, S., University of Minnesota
Xu, F., Beijing Forestry University
Originally presented on: 11/6/2013 16:05:00 - 16:30:00

Lignocellulosic biomass consists of three major compounds including cellulose, lignin and hemicellulose and small amounts of extractives comprising approximately 90% of the dry weight of these plant materials. Each of these four components has a different degree of resistance to chemical, thermal and biological degradation.4 The chemical and biochemical structure has been perfected by nature to resist degradation and maintain the integrity of biomass. Pretreatments and conversion techniques deconstruct the complex plant cell wall to release the components of the biomass ([i], [ii], [iii], [iv]). Acid, alkaline and enzymatic approaches show varying yields and degrees of success of biomass conversion ([7, 12]). There has been very little effort in understanding the chemical and morphological characteristics of secondary cell walls and how they are influenced by biomass conversion strategies ([v], [vi]). Currently available pretreatment techniques used in the conversion of biomass to chemicals and energy provide a broad range of choices that improve the conversion of carbohydrates from lignocellulose biomass to biofuels, biochemicals, and biomaterials. In general, these techniques focus on disrupting the protection of cellulose and hemicelluloses by lignin, as well as weakening the dissociation between these constituents. While these pretreatment methods are adjusted to the structure and properties of the biomass raw material chosen, they neither provide adequate yields of fermentable glucose from cellulose, nor are economically sound or energy efficient. To overcome the tradeoff between high yield and energy usage research must clarify the chemical pathways which constituents undergo during the pretreatment methods. Understanding of the ultrastructure and topochemistry of the cell wall constituents during pretreatments will lead to the improvement of the yields of fermentable sugars while reducing the energy consumption improving the cost model of production to these alternative bioproducts. The efficacy of enzymatic complexes to hydrolyze these substrates is inextricably linked to the innate structural characteristics of the substrate and/or the modifications that occur as saccharification proceeds.

A discussion of the application of new techniques including immunoassay labeled nanoparticle tags, correlative imaging with AFM, nanoCT and Raman will be given. Representative examples of the applications to biomass samples is also presented.

[i] Donohoe, B., Decler, S., Tucker, M., Himmel, M., Vinzant, T., Biotechnology Bioengineering, 2008,

101, 913-925.

[ii] Kremer, J., Mastronarde, D., McIntosh, J., Journal of Structural Biology, 1996, 116, 71-76.

[iii] Mastronarde, D.N., Journal of Structural Biology, 1997, 120, 343-352.

[iv] C. E. Wyman, B. E. Dale, R. T. Elander, M. T. Holtzapple, M. R. Ladisch and Y. Y. Lee, Bioresour. Technol., 2005, 96, 1959–1966.

[v] J. Nakashima, T. Mizuno, K. Takabe, M. Fujita and H. Saiki, Plant Cell Physiol., 1997, 38, 818–827.

[vi] Somerville, C., Bauer, S., Brininstool, G., Facette, M., Hamann, T., Milne, J., Osborne, E., Paredez, A., Persson, S., Raab, T., Vorwerk, S., Youngs, H. Science, 2004, 306, 2206–2211.

(580d) Full Components Analysis and Directional Control Research of Wood Powder Pyrolysis Products

Chen, L., Energy Research Institute of Shandong Academy of Sciences
Zhang, X., Energy Research Institute of Shandong Academy of Sciences
Meng, M.,
Zhao, B., Energy Research Institute of Shandong Academy of Sciences
Sun, L.,
Si, H.,
Originally presented on: 11/6/2013 16:30:00 - 16:55:00

Fast pyrolysis of wood power was performed and the pyrolysis mechanism was researched. The condensable and noncondensable pyrolytic gases were analyzed with on-line pyrolysis-gas chromatography/mass spectrometry (PY-GC/MS) and Agilent GC6890N, the solid produces were analyzed with elemental analyzer. It was investigated that the effect of pyrolysis temperature, resident time and catalysts on the pyrolysis components, and the directional controlling measures for the produces distribution were achieved.

(580e) Biogas Production from Lipid-Extracted Algal Residue

Li, Y.,
Zhang, X., Energy Research Institute of Shandong Academy of Sciences
Hua, D.,
Zhang, J., Shandong University
Zhao, Y.,
Xu, H.,
Liang, X.,
Mu, H.,
Originally presented on: 11/6/2013 16:55:00 - 17:20:00

Waste-grown microalgae is a potentially important biomass for wastewater treatment, meanwhile lipid extracted from the algae could be further converted into biofuels. The lipid-extracted algal residues as major by-products mainly consist of carbohydrate and protein, which makes anaerobic digestion an efficient way to recover energy. The conversion of lipid-extracted algal residues into methane plays dual role in renewable energy production and sustainable development of microalgal biodiesel industry. Therefore, an anaerobic fermentation process for investigation of methane production potential of algal residues was conducted in this paper. The effect of inoculum to substrate ratios (ISRs) on the methane production by anaerobic digestion of *Chlorella sp.* residue in single stage was evaluated. The stability and progress of the reaction from algal residues to methane were monitored by measuring the pH, volatile fatty acids (VFAs), total ammoniacal nitrogen (TAN), methane volume. Due to the characteristic of high content of protein in algal residues, the inhibition would occur resulting from the generation of ammonia. Thus two stage technology was proposed and more suitable to improve the methane yield.

(580f) Study on Pyrolysis Properties of Agriculture Residues by Using FTIR and DRIFT Technique

Fu, P., Shandong University of Technology
Yi, W., Shandong University of Technology
Li, Z., Shandong University of Technology
Bai, X., Shandong University of Technology
Li, Y.,
Wang, J., Shandong University of Technology
Originally presented on: 11/6/2013 17:20:00 - 17:45:00

Study on pyrolysis properties of agriculture residues by using FTIR and DRIFT technique

Peng Fu, Weiming Yi*, Zhihe Li*, Xueyuan Bai, Yanmei Li, Jing Wang

School of Agricultural and Food Engineering, Shandong University of Technology, Zibo 255049, China

*Corresponding author. Tel.: +86 533 2786308; Fax: +86 533 2786158

Abstract

The gas release properties and char structural evolution during pyrolysis of agricultural residues (maize stalk, rice straw and cotton straw) was studied with Fourier transform infrared spectroscopy (FTIR) and diffuse reflectance infrared Fourier transform spectroscopy (DRIFT).

The results showed that the pyrolysis processes underwent three consecutive stages, corresponding to the evaporation of water, the formation of primary volatiles and the subsequent release of small molecular gases. Fig. 1 showed that the major pyrolysis gases for the three materials were similar, including CO_2, CO, methane, ethane, and some organics such as methanol, formaldehyde, formic acid, acetic acid and acetone. HCN was the major nitrogen containing product. The release of CO_2 mostly focused in the temperature range of 250?450 °C. It was mainly released out with the cracking and reforming of carbonyl and carboxyl groups. The evolution of CO started at ≈ 230 °C, increased significantly with rising temperature, and reached a maximum at 337, 301 and 339 °C for maize stalk, rice straw and cotton straw respectively. The formation of CO below 400 °C was mainly caused by the cracking and reforming of thermolabile carbonyl and ether groups. The release of CO above 450 °C was probably due to the scissions of diaryl ether groups and secondary reactions of decomposition of the char. Compared with CO and CO_2, CH_4 evolution covered over a wider temperature range of 300?600 °C, with a maximum at about 420 °C. The formation of methane below 500 ^{o}C was mainly attributed to the cracking of the methoxyl groups. Moreover, the breaking of methylene groups in the temperature range could also contribute to the formation of CH_4. The release of methane at higher temperatures might be ascribed to the rupture of aromatic rings. The aliphatic ?CH_2OH groups in ?γ position of the alkyl side chains and aromatic methoxyl groups were the main source of methanol. The reforming of free hydroxyl groups and C?O groups could also generate methanol. The formation of formaldehyde at lower temperatures indicated that the aliphatic ?CH_2OH groups might be easily removed by alkyl C?C fragmentation.

To investigate the changes in the surface chemistry of chars formed during biomass pyrolysis, information on the surface chemistry of raw samples and chars has been provided by DRIFT spectroscopy. The DRIFT spectra of rice straw and chars given in Fig. 2 showed that the amount of hydroxyl (3200?3700 cm^{-1}), aliphatic C-H groups (2800?3000 cm^{-1}) and olefinic C=C bonds (1610?1680 cm^{-1}) in the chars decreased significantly when the temperature exceeded 250 ^{o}C. The amount of C=O (1650?1770 cm^{-1}) and ether structures in the chars first gradually decreases and then increases with the temperature rising. The aromatic structure develops when the temperature was higher than 350 ^{o}C. Aromatization was also evidenced by the shift to lower frequencies of the bands due to *ν*(C=O) and olefinic *ν*(C=C) vibrations. This behavior was consistent with the increase in conjugation of C=O groups and C=C bonds. At 500 ^{o}C, the char contained large amounts of aromatic and ether type structures. With the further increase of temperature, the amount of ether groups in the char decreases and the char became increasingly polyaromatic. The loss of ether groups led to a more ordered carbon structure at high temperature. The DRIFT analysis showed that the hydroxyl, aliphatic C-H, carbonyl and olefinic C=C functional groups were lost at high temperatures. The aromatization process started at ≈ 400 ^{o}C and continued to higher temperatures.

Keywords: Agricultural residues; Pyrolysis; FTIR; DRIFT

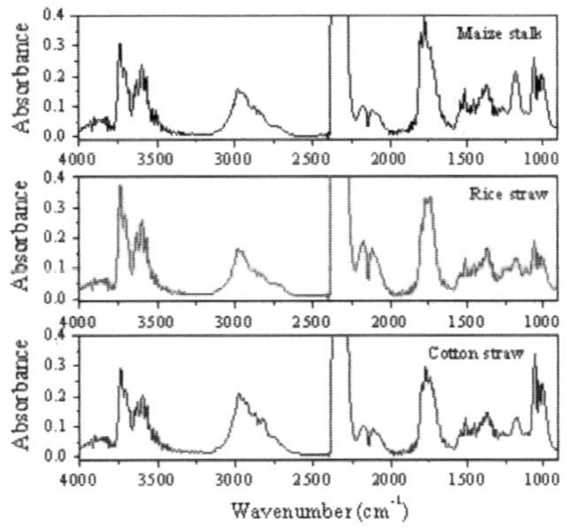

Fig. 1. FTIR spectra of gas products evolved at 300 ^{o}C for pyrolysis of three materials.

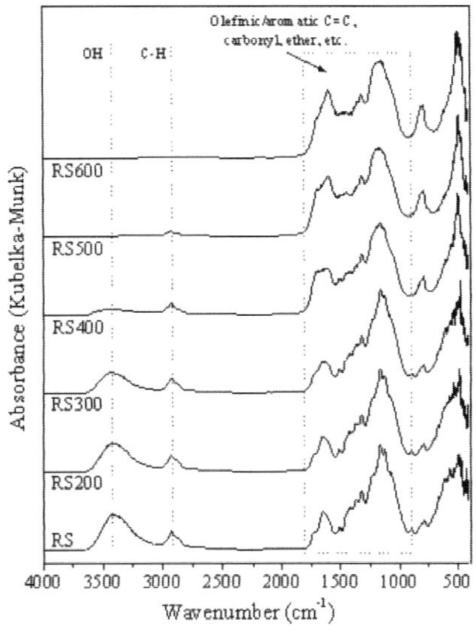

Fig. 2. DRIFT spectra of rice straw and chars formed at different temperatures.

(599a) Hydrodeoxygenation of Microalgae Oil in a Microreactor

Zhou, L.,
Lawal, A., Stevens Institute of Technology
Originally presented on: 11/7/2013 8:30:00 - 8:55:00

Removal of residual oxygen from long chain lipids via hydrodeoxygenation (HDO) will produce a liquid fuel comprising mostly straight chain alkanes, which mimics diesel. Green diesel is chemically similar to petroleum-derived diesel, and it can be used in existing internal combustion engines without any modification. Microalgae has been favored as a feedstock for green diesel, on account of its high acylglycerols contents and faster growth rate than terrestrial oil crops. Moreover, HDO, a process that uses high pressure hydrogen to remove oxygen is CO_2 neutral.

In the present work, sulfided $NiMo/Al_2O_3$ catalyst was evaluated for HDO with the aim of converting algae oil to diesel-like hydrocarbons. Microreactor was used due to its advantages compared to conventional reactors. These advantages include: 1) enhanced mass and heat transfer, 2) less energy consumption, 3) faster and cost-saving screening of materials and process, 4) easier scale-up, and 5) improved safety. Changes in the hydrogen consumption and the product composition as a function of operating conditions: reaction temperature, hydrogen partial pressure, residence time were determined quantitatively by GC-TCD and FID.

This work will provide baseline data which will guide future work on the HDO of algae oil.

(599b) Xylulose to Furfural, a New Route to Produce High Yield of Furan

Li, B., University of Toledo
Marbaugh, K., University of Toledo
Relue, P., University of Toledo
Varanasi, S., University of Toledo
Originally presented on: 11/7/2013 8:55:00 - 9:20:00

Biomass has the potential to serve as a renewable feedstock for the production of a wide range of fuels and chemicals. Furfural, which is currently made from waste biomass, is a very versatile and key platform chemical for the synthesis of, among others, 2-methylfuran, gasoline and diesel components. Current production methods are limited in the efficiency of furan production due to incomplete conversion of the xylose as well as undesired byproduct and humin formation. In an effort to increase the yield and selectivity of furfural production, we have produced furfural from xylulose, the ketose isomer of xylose. Production of furfural directly from xylulose has not been considered a viable option due to the excessive cost of high purity xylulose. However, we have developed a method of simultaneous isomerization and reactive extraction (SIRE) for producing concentrated xylulose at high purity. We have evaluated several potential reaction models to determine the reaction kinetics of the xylulose dehydration and side reactions that affect furfural yield in different reaction media. By comparing the predictions of the models to experimental data on xylulose dehydration, we have been able to determine the most probable reaction mechanisms as well as optimal conditions for furfural production from xylulose.

(599c) Heterogeneous Catalytic Oxidation of Biorefinery Lignin for Dicarboxylic Acid Production

Ma, R.,
Zhang, X., Bioproducts Science Engineering Laboratory, Washington State University
Originally presented on: 11/7/2013 9:20:00 - 9:45:00

Directly production of dicarboxylic acid from biorefinery lignin is introduced as one new direction of lignin conversion to value-added chemicals, especially dicarboxylic acids. A mild condition oxidation of lignin phenolic units over catalyst is suggested to be a promising pathways to form ring-opened products, such as muconic acid, malonic acid, and succinic acid. Lignin is the largest volume renewable feedstock that contains aromatic rings, whose application is nowadays still limited to thermovalorisation processes. Its potential as a source of valuable phenolic in the production of value-added chemicals in alternative to petrol chemistry is largely unexpoloited. In contrast to most previous literatures, industrial and biorefinery lignin, instead of lignin model compounds, were tested in this study. We also focused on developing a green conversion technology which eliminates the use of hazardous chemicals, harsh reaction condition and high capital-cost equipment. Oxidative catalytic process provides alternative and perhaps much more economically viable route compared to reduction pathways to convert lignin to precursors for hydrocarbon fuel and bioproducts production.

In this study, we introduce an aqueous phase oxidation of various biorefinery lignins over a class of chalcopyrite catalysts to directly produce dicarboxylic acid. A clear combination of two stage reactions, including depolymerization and oxidative aromatic ring cleavage, is observed. For example, diluted acid treated corn stover is depolymerized into mixture of lignin monomeric fragments, including benzoic acid derivatives, cinnamic acid derivatives, and hydroquinone. In the second stage, a mixture of succinic acid, malonic acid, maleic acid and malic acid, makes up to 87% in final effluence determined by GC-MS. Dimer and monomer study of gives the similar results which confirmed the occurrence of aromatic ring-cleavage reactions as the source of dicarboxylic acid generation. With deeper understanding of reaction mechanisms, we suggest that various lignin units lead to different reaction pathways, and also results. P-hydroxyphenyl units, in which hydroquinone derivatives work as the intermediates, are likely to form maleic acid and malonic acid from the ring cleavage of p-quinone structure. In another way, guaiacyl units tend to produce maleic acid and succinic acid as the final products mediated by the formation of muconic acid as the precursor from ring-opened mechanism. In conclusion, we believe the chalcopyrite structured catalysts could selectively convert lignin into dicarboxylic acid in a ring-opened mechanism.

(599d) Hydrothermal Carbonization: Reactions and Water Production

Uddin, M. H., University of Nevada, Reno
Reza, M. T., University of Nevada, Reno
Lynam, J. G., University of Nevada, Reno
Hoekman, S. K., Desert Research Institute
Coronella, C. J., University of Nevada, Reno
Originally presented on: 11/7/2013 9:45:00 - 10:10:00

Hydrothermal carbonization (HTC) is a promising thermochemical conversion process for lignocellulosic biomass to offer value-added products. Biomass is not required to be dry, as the treatment is actually conducted in hot water. The process is done at temperatures between 200 °C and 280 °C, and the pressure is high enough to maintain liquid water. HTC converts lignocellulosic biomass to solid bio-carbon, aqueous phase chemicals, and gases.

In the decomposition of lignocellulosic biomass, knowledge of the product yields and water production is very important. In this study chemical reactions occurring during the relatively short HTC reaction time are discussed. Analytical techniques include characterization of solid and liquid HTC products through ATR/FTIR, ultimate analysis, GC-MS, and IC. HTC reactions for loblolly pine are proposed in the context of HTC reactions for individual biomass components. Based on the results, dehydration and decarboxylation reactions are proposed to be the major reactions of HTC after hydrolysis, though condensation, polymerization, and aromatization also occur simultaneously.

An experimental procedure was developed to quantify dehydration rate and the production or consumption of water. Experiments were performed with loblolly pine at 200 °C, 230 °C, and 260 °C for 5, 15, and 30 min reaction times. For the first time, we present quantitative results showing yield of water by dehydration and hydrolysis reactions. Detailed analysis shows that water used for initial hydrolysis at 200 °C cannot be recovered with a 30 min reaction time. However, dehydration and water production increases with both HTC temperature and reaction time. With longer reaction times, the conversion rate of biomass to high carbon content char apparently slows down within 5 to 10 min, with side reactions occurring in the liquid phase during the rest of the reaction time. In general, the solid phase hydrolysis reactions are much faster than the aqueous phase dehydration reactions.

(599e) A Fresh Look At The H Factor DATA For Indian Nonwood Agri-Residues

Ray, A. K., Indian institute of technology, Roorkee
Ravimohan, R.,
Prasad, R.,
Yadav, R.,
Originally presented on: 11/7/2013 10:10:00 - 10:35:00

A FRESH LOOK AT THE H FACTOR DATA FOR INDIAN NONWOOD AGRI-RESIDUES

ABSTRACT

H factor is an extremely important parameter in modern process control in pulp mill operation representing extent of chemical reaction as it enables changes to be made in the time- temperature cycle while maintaining a constant degree of delignification. The combination of time and temperature as a single variable does not have any effect as long as the H factor remains constant, the yield and lignin content will be the same. The greater the H Factor more the delignification occurs. Generally, H factor is determined using the procedure proposed by Vroom , known as Vroom H factor model which provides a method for expressing both cooking time and temperature in H factor for soft wood. There have many modifications of this parameter and its application during pulping studies over the years. As per actual definition,it should be the integral of the relative reaction rate or the absolute reaction rate with respect to time. The H-Factor values are thus computed based on both mathematical methods using graphical and numerical integration techniques as well as approximate method based on Vroom's H-Factor used for routine control..

For many nonwood annual fibres such as wheat straw, rice straw and bagasse , the H factor values have not been established even today. Attempts have made in this paper to model the H factor for some important annual fibers ,namely bagasse and straw which are very important feedstock for pulp and paper industries in India. Attempts have also been made to highlight the various modifications of H factor estimation procedure and compare the values. As for estimation of H factor activation energy values are required, this needs in turn delignification reaction to be performed. Therefore in the present work detailed kinetic experiments are carried out and activation energy for these fibres is determined. Using the experimentally evaluated value of activation energy, Boltzmann frequency factor ,and specific reaction rate kinetic model for delignification is developed. A comparison has been made for the values of absolute H-Factor ,Vroom's H-Factor , Jagannadh et al. ,others based on developed kinetic modals and H-Factor calculated using MacDonald's data for wood is made.

(663a) Study of the Hydrolysis of Rice Straw Using Phosphoric Acid,Amberlite Catalyst and Process Optimization Using RSM Methodology

Pratap Yadav, S., Iit Roorkee India
Ray, A. K., Indian institute of technology, Roorkee
Ghosh, U. K., IIT Roorkee
Originally presented on: 11/7/2013 12:30:00 - 12:55:00

Agricultural wastes are renewable, cheap and easily available sources of value added chemicals. Due to depletion of nonrenewable sources of energy, it is required to find out alternative ways to produce value added chemicals which are produced by petroleum refing process as byproducts. In this work, samples of milled rice straw were hydrolyzed with phosphoric acid in presence of amberlite catalyst under varying conditions of reaction time, reaction temperature, solid to liquid ratio and wt% of catalyst to find out the optimum process condition to produce different sugars(xylose, arabinose, glucose) and decomposition products(furfural, hydroxymethyl furfural). Remaining solid mass, sugar concentrations and decomposition product concentrations were measured. For measurement HPLC technique was used. Different kinetic models were developed to describe the kinetics of products of reaction in varying reaction conditions . Different kinetic parameters were find out using regression method.These models were also compared with available models in the literature. Using experimental data, the process conditions were optimized by RSM methodology.

(663b) Furfural Production As a By-Product of Lignocellulosic Biofuel Process

Chen, X., National Renewable Energy Laboratory
Nagle, N. J., National Renewable Energy Laboratory
Tao, L., National Renewable Energy Laboratory
Tucker, M., National Renewable Energy Laboratory

Originally presented on: 11/7/2013 12:55:00 - 13:20:00

NREL's high solids dilute acid steam pretreatment often produces a xylose rich stream with a monomeric xylose concentration in the range of 70~100g/L. The high xylose concentration inhibits the digestibility of biomass during the subsequent whole slurry enzymatic hydrolysis step, leading to a decrease in both glucose and xylose yields. In addition, xylose is not a fermentation-friendly sugar, especially to hydrocarbon fuels. However, xylose can be converted into furfural through dehydration reactions using acid catalysts. At the present time furfural is a high-valued commodity product, where in 2010, the US market average price of furfural was $8.78/gal, over 3 times higher than ethanol's price. In addition, Furfural is a promising intermediate in the production of hydrocarbons and other chemicals. In this study, alternative pathways to produce furfural as a byproduct of lignocellulosic bioethanol processes were investigated. A delayed decompression flash process was adopted to effectively produce furfural while reducing the degradation of cellulose and furfural to other degradation products. The more severe reaction conditions also significantly reduce the enzyme requirements and improve the digestibility of the residual cellulose. We also evaluated the fermentability of the sugar steam after enzymatic hydrolysis. . This work provided preliminary experimental results to support a techno-economic analysis (TEA) completed using an Aspen plus model modified from the 2011 NREL design base case.

(663c) A Kinetic Interpretation of Autohydrolysis of Mixed Southern Hardwood Chips in a Pilot Scale Reactor

Yan, J., SUNY ESF
Xie, Y., SUNY ESF
Huang, C., Guangxi University
Liu, S., SUNY College of Environmental Science and Forestry
Originally presented on: 11/7/2013 13:20:00 - 13:45:00

Woody biomass is an important sustainable carrier of chemicals, materials and energy for humanity. Selectively extracting five carbon sugars from woody biomass is an intelligent choice for bioconversion of woody biomass. Mixed southern hardwood chips were loaded in a pilot-scale reactor of 1.84 m^3. Tap water was filled into the reactor after woodchips were loaded. The reactor contents were then heated by circulating the water in the reactor with a heat-exchanger using steam as heating medium. As temperature was raised, liquid samples were also collected. Liquor pH decreases as reaction time is extended reaching a plateau of about 3.5. Xylose, glucose, mannose, galactose, arabinose and rhamnose are found in the liquor along with their oligomers. The reaction can be described by a first order kinetics. Odexlims was utilized to evaluate the kinetic parameters for this complex non-isothermal reaction process.

(663d) Mechanism Study of Noble Metal Catalyzed Upgrading of Lignin Derived Pyrolysis Oil

Mu, W., Georgia Institute of Technology
Ben, H., National Renewable Energy Laboratory
Ragauskas, A. J., Georgia Institute of Technology
Deng, Y., Georgia Institute of Technology
Originally presented on: 11/7/2013 13:45:00 - 14:10:00

Lignin is the second most abundant natural polymer in the world. However, unlike cellulose and hemicellulose, over 98% of lignin is simply burned for heat energy. Pyrolysis is recognized as one of the most economical ways to convert solid lignin to liquid oil. However, the oil products from this process cannot be directly used as a fuel. The major problems are instability, high acidity and low heating value. Therefore, upgrading the lignin pyrolysis oil is one of important steps in converting lignin to biofuel. In this research, palladium, platinum, rhodium and ruthenium were used as active metal in catalyst. Palladium and platinum showed no reactivity with the pyrolysis oil. Ruthenium was the only catalyst which is able to produce fully hydrogenated and intensively deoxygenated products. Guaiacol was adopted as model compound to study the reaction mechanism of the oil upgrading. The result revealed that palladium and platinum were incapable of both hydrogenation and deoxygenation reaction with guaiacol due to the deactivation of catalyst. On the other hand, ruthenium was able to achieve both reactions. Related kinetics study showed different reaction path and reactivity for ruthenium and rhodium with the model compound.

(663e) 5-Hydroxymethylfuran and 2,5-Dimethylfuran Production From Banana Waste

Sierra-Salazar, A. F., Universidad Nacional de Colombia
Torres-Conde, A., Universidad Nacional de Colombia
Ruiz-Colorado, A., Universidad Nacional de Colombia
Originally presented on: 11/7/2013 14:10:00 - 14:35:00

5-Hydroxymethylfuran and 2,5-Dimethylfuran Production from Banana Waste

Andres Felipe Sierra-Salazar[1], Andres Torres-Conde[2], Ángela Adriana Ruiz-Colorado[3]

[1] afsierrasa@unal.edu.co, [2] atorresc@unal.edu.co, [3] aaruiz@unal.edu.co

Grupo de Bioprocesos y Flujos Reactivos

Facultad de Minas

Universidad Nacional de Colombia – Sede Medellín

Keywords: 2,5-dimethylfuran, 5-hydroxymethylfurfural, hydrolysis, hydrogenolysis, banana.

2,5-Dimethylfuran (DMF), C_6H_8O, has interesting chemical and physical properties: high energy density (31.5 MJ/L, compared to gasoline: 35.0 MJ/L, and ethanol: 23.0 MJ/L, (Tong, Ma, & Li, 2010)), high boiling point (20K higher than ethanol) and little solubility in water (Román-Leshkov, Barrett, Liu, & Dumesic, 2007). Indeed, DMF has the lowest water solubility and the highest research octane number (RON) among all the mono-oxygenated C6 compounds. These observations show the potential of DMF as an additive or alternative of transportation fuels (Tong et al., 2010).

DMF can be produced by 5-hydroxymethylfurfural (HMF) catalytic hydrogenolysis (Tong et al., 2010). As intermediate, 2,5-bis(hydroxymethyl)furan (BHF) is also produced. The latter can be used in the industry of polymers, resins and artificial fibers (Lewkowski, 2001).

Banana: source of hexoses

Just in Colombia, the surplus of banana production is around 120.000 tons per year, this biomass is then dropped. In other banana producer countries the surpluses of fruit also exist due to damage, inadequate maturity, overproduction, etc. About 70% of banana world exports are form Ecuador, Philippines, Costa Rica, and Colombia (Correa Uribe, 2011).

Banana fruit is primarily composed of starch and its peal, of cellulose. The shoot and pseudo-stalk (other wastes from banana culture) have also important amount of cellulose. These compounds can be hydrolyzed to glucose. Traditionally, the sugary liquor produced by hydrolysis is usually fermented to ethanol (Correa Uribe, 2011). However, glucose monomers are hexoses that could be converted to HMF; an alternative of adding value to wastes and surplus in the banana production chain.

5-Hydroxymethylfurfural (HMF)

HMF can be obtained from hexoses, polysaccharides and industrial wastes. From hexoses, the occurring reaction is a triple dehydration. However, the obtaining of this compound is not an easy task because the HMF is an intermediate compound in the reaction from sugars to levunilic and formic acids. Also, there is the possibility that the HMF polymerize into humic acids at high concentrations (Lewkowski, 2001).

In addition, as stated before, the production of HMF is carried out through a dehydration process, which is more complex when it is considered that lignocellulosic residuals besides being composed by cellulose, hemicellulose and lignin, are materials rich in water (Lewkowski, 2001). All these limitations open the door for an optimization development in the HMF production.

Experimental Procedure

The lignocellulosic material was homogenized by milling and mixing. Then, moisture was determined using the NREL method *TP-510-42621: Determination of Total Solids in Biomass and Total Dissolved Solids in Liquid Process Samples* (Sluiter et al., 2008). After, the ratio dry-solid / liquid and acid concentration were adjusted for the hydrolysis. The reactions were carried out at 100rpm.

For the experimental design defined to optimize the production of HMF, three variables were selected to be handled: time, temperature and acid concentration. These have the biggest impact on the obtaining of HMF (Correa Uribe, 2011). Previous experiments were also carried out using these factors in a reduced factorial 3^{3-1}design. From that, the ranges of higher HMF concentrations were achieved without sulfuric acid, temperatures above 150°C and for about 4 hours (Sierra-Salazar & Ruiz-Colorado, n.d.).

Having those results on mind, the factors selected for acid, concentrations of 0.0 and 0.5 % w/w, for temperature 165°C and 210°C, for time were 2 and 4 hours, since with higher temperatures, the time is expected to be reduced. The chosen design was a response surface - central composite design unblocked with 20 experiments including 6 repetitions on the central point.

After the hydrolysis reaction, liquid extraction was performed in 4 successive stages with chloroform. Then, a small amount of high boiling point solvent was added to the organic phase in order to proceed with the evaporation of chloroform using vacuum rotary equipment. HMF was finally obtained in the desired solvent (Rufian-Henares, Garcia-Villanova, & Guerra-Hernandez, 2001).

The hydrogenolysis reaction to DMF was carried in two different solvent systems: water and water - n-butanol. The reaction conditions were: 20ml of medium volume, 0.4g of catalyst, 4.0bar as initial hydrogen pressure and 220°C for 10 hours (Román-Leshkov et al., 2007).

The catalysts for this reaction were copper supported on silica or on γ-alumina, with barium as a promoter in some trials. For their preparation, the incipient impregnation method was used (Guo, Zhou, Mao, Guo, & Zhang, 2009), which includes the deposition of the catalytic compounds ($CuNO_3$, $BaNO_3$) by dropping them in solution over the supporting material. The solvent for these solutions has to be chosen considering the wetting angle relative to the support, in order to be adsorbed onto the carrier surface (Yang, 2003).

After impregnation, the catalyst were dried for 12 hours at 110 ° C and thereafter calcined at 400 ° C for 3 hours and finally the catalyst was reduced *in situ*in the reactor under the reaction conditions. The copper and barium ratio are 2.7mmol/g support and 0.3mmol/g support, respectively (Guo et al., 2009).

Results

HMF was detected by HPLC with PDA (SPD-M20A, Diode Array Detector) and levulinic acid with RID-10A (Refrective Index Detector), from Shimadzu. The conditions included a column: Aminex HPX-87H Bio-Rad, mobile phase $H_2SO_4$5mM at 60°C, 0.6ml/min and 20uL of injection volume.

After an ANOVA analysis, the maximum production of HMPF was found to be 3.08% g HMF / g initial biomass, keeping close to zero the concentration of levulinic acid. This value was obtained under the conditions of almost non sulfuric acid and around 190°C. For a scaling of the process, we recommend a series of reactors and between them, separation steps, in order to convert the maximum amount of biomass to 5-hydroxymethylfurfural.

For the conversion of HMF to DMF in water, four catalysts were used, two supported on commercial silica and two on alumina. The highest conversion was obtained with the supported on alumina (~ 100%) compared with copper on silica (96.7%) and mixture of copper and barium on silica (95.5%). For the production of DMF in a biphasic system (50% water – 50% n-butanol), the same catalysts were used and the conversion of HMF obtained with all of them was around 81%.

According to these results, Cu/Silica and Cu/γ-Alumina did not show significant differences in reactive conversions.

References

Correa Uribe, J. F. (2011). Identificación y modelamiento de la producción de furfural e HMF en el proceso de hidrólisis ácida de excedentes de la fruta de banano. Universidad Nacional de Colombia.

Guo, L., Zhou, J., Mao, J., Guo, X., & Zhang, S. (2009). Supported Cu catalysts for the selective hydrogenolysis of glycerol to propanediols. Applied Catalysis A: General, 367(1-2), 93–98. doi:10.1016/j.apcata.2009.07.040

Lewkowski, J. (2001). Synthesis , chemistry and applications of 5-hydroxymethylfurfural and its derivatives. Arkivoc, 2001(i), 17–54.

Román-Leshkov, Y., Barrett, C. J., Liu, Z. Y., & Dumesic, J. a. (2007). Production of dimethylfuran for liquid fuels from biomass-derived carbohydrates. Nature, 447(7147), 982–5. doi:10.1038/nature05923

Rufian-Henares, J. A., Garcia-Villanova, B., & Guerra-Hernandez, E. (2001). Determination of Furfural Compounds in Enteral Formula. Journal of Liquid Chromatography & Related Technologies, 24(19), 3049–3061. doi:10.1081/JLC-100107356

Sierra-Salazar, A. F., & Ruiz-Colorado, Á. A. (n.d.). 5-hydroxymethylfurfural and 2,5-dimethylfuran production from banana waste.

Sluiter, A., Hames, B., Hyman, D., Payne, C., Ruiz, R., Scarlata, C., Sluiter, J., et al. (2008). Determination of Total Solids in Biomass and Total Dissolved Solids in Liquid Process Samples Biomass and Total Dissolved Solids in Liquid Process Samples, (March).

Tong, X., Ma, Y., & Li, Y. (2010). Biomass into chemicals: Conversion of sugars to furan derivatives by catalytic processes. Applied Catalysis A: General, 385(1-2), 1–13. doi:10.1016/j.apcata.2010.06.049

Yang, R. T. (2003). Adsorbents fundamentals and applications. Chemical Engineering.

(716a) Bioprospecting the Cellulases and Xylanases Thermozymes for the Production of Biofuels

Kumar, S., Sardar Swaran Singh National Institute of Renewable Energy
Behera, S.,
Arora, R.,
Originally presented on: 11/7/2013 15:15:00 - 15:40:00

Bioprospecting the cellulases and xylanases thermozymes for the production of biofuels

Shuvashish Behera, Richa Arora, Sachin Kumar*

Sardar Swaran Singh National Institute of Renewable Energy, Jalandhar-Kapurthala Road, Wadala Kalan, Kapurthala, Punjab-144601, India

Abstract

A wide variety of microorganisms are present in the environment, which can be harnessed for the purpose of mankind. The isolation and screening of the efficient enzymes like cellulases and xylanases producing microbes can be helpful to overcome the current challenges in biofuel production. This study focuses on isolation of efficient cellulase and xylanase producing thermophilic bacteria found in different soil samples, which could be used in enzymatic saccharification of biomasses. The bacterial isolates with appreciable hydrolytic zones were isolated from soil sample based on zone of clearance on solid plates. All isolates were studied by cultural morphological characteristics and microscopic examinations. About 17 bacterial strains showed the cellulose and/or xylanase enzyme activity. Some isolates showed individual enzyme activity i.e. either cellulase or xylanase and rest showed the activity for both the enzymes. Four isolates designated NIRE C3, NIRE C16, NIRE U1 and NIRE U2 showed superior cellulase activity with more than 15 mm zone of clearance and two isolates NIRE B7 and NIRE B8 shows highest xylanase activity producing more than 16 mm zone of clearance after 24 h of incubation at 50°C. These cultures represent excellent candidates for further enzyme production and characterization.

*Corresponding author. Tel.: +91 1822 255543; fax: +91 1822 255544.

E-mail address: sachin.biotech@gmail.com (S. Kumar)

(716b) Novel Thermostable Xylanase Production From Termite Gut Microbes and Its Application in Biofuels

Dheeran, P., 1SRM Research Institute, SRM University
Kumar, S., Sardar Swaran Singh National Institute of Renewable Energy
Adhikari, D., Indian Institute of Petroleum
Originally presented on: 11/7/2013 15:40:00 - 16:05:00

Novel thermostable xylanase production from termite gut microbes and its application in biofuels

Pratibha Dheeran[1,*] , Sachin Kumar2 and Dilip K. Adhikari[3]

[1]SRM Research Institute, SRM University, Kattankulathur-603 203, Tamilnadu, India

[2]Sardar Swaran Singh National Institute of Renewable Energy, Kapurthala-144601, India

[3]Biotechnology Area, Indian Institute of Petroleum, Dehradun-248 005, India

Abstract

The termites harbor more than 200 species of microbes that produce the novel cellulosic and hemicellulosic degrading enzymes, which in turn can be efficiently applied for the production of ethanol from lignocellulosic biomass in biofuel industries. In the current work, wood feeding termites were collected from the forest of Rajaji National Park Range, Dehradun (India) and they were used to isolate the thermostable xylanse producing microbes. As far as literature concern, very few information is available on using the termite gut microbes for xylanase production.

The wood feeding termite gut's extract were used for screening of xylanase producing microbes at 50°C. A novel thermophilic strain *Panebacillus macerans* IIPSP3 was isolated which was found to be potent thermostable xylanase producer. Maximum xylanase production was observed in late exponential phase at 50°C and pH 6.0. The isolated thermostable xylanase worked at broad temperature and pH range 40-90°C and 3.5-9.5, respectively, with optimum activity at temperature of 60°C and pH 4.5. This novel strain *P. macerans* IIPSP3 from wood feeding termites could utilize agro wastes such as corn cob as a substrate for growth and produce high level of xylanase, which has benefits for its economical production and industrial applications. IIPSP3 xylanase is a novel enzyme, which can be used in biofuel and paper & pulp industries as it could efficiently be hydrolyzed the lignocellulosic biomass at pH 4.5 and 9.0. The isolated xylan is cellulase free, which can be used in biobleaching of pulp.

Key Words: Higher Termite, Thermostable xylanase, Biofuels.

*Author for correspondence: E-mail: pratibhadheeran@gmail.com

(716c) Inducing Cellulase Activity of Microbulbifer Hydrolyticus Using Hot-Water Wood Extract As Substrate

Arens, K., SUNY College of Environmental Science and Forestry
Liu, S., SUNY College of Environmental Science and Forestry
Originally presented on: 11/7/2013 16:05:00 - 16:30:00

This study used wood extract from *Paulownia elongata* (WE) and hydrolisate of sugar maple (hydrolysate) produced during hot-water extraction as substrates to stimulate the production of cellulase from *Microbulbifer hydrolyticus* IRE-31(ATCC 700072). Selective strain medium adaptation of the original wild *M. hydrolyticus* IRE31 was performed in order to improve bacterial growth in mediums supplemented with hydrolysate and WE. Comparative studies were performed to establish better bacterial growth conditions and enzymatic activity at a given temperature, agitation and medium pH. The effect of different substrates, growth supplements and bases utilized to control the medium pH were among the studies evaluated. Flask tests where pure marine broth medium with an enzymatic activity of 0.009 U/ml broth medium was used as control and compared it to mediums supplemented with xylose, glucose, cellobiose, hydrolisate, WE or yeast extract, showed higher cellulase activity in the sample broth containing either WE or hydrolisate. No discernible activity was revealed in broths containing either glucose or xylose. Scale up from 100ml flask incubations to 1L stirred tank bioreactor set at pH 7.5, 37Â¢Â°C and 300 rpm, and supplemented with 15% (v/v) *P. elongata* wood extract or 10% (v/v) sugar maple hydrolisate, yield a maximum enzymatic activity of 1.417 U/ml broth medium and 0.678U/ml broth medium respectively. These results reveal a significant increase in cellulase activity with growth mediums supplemented with complex substrates containing xylooligosaccharides such as *P. elongata* wood extract and sugar maple hydrolysate.

(716d) Synthesis and Performance of Cellulase-Mimetic Polymeric Solid Catalysts for Hydrolysis of Cellulose

Yang, Q., University of Wisconsin, Madison
Pan, X., University of Wisconsin, Madison
Originally presented on: 11/7/2013 16:30:00 - 16:55:00

Hydrolysis of cellulose to glucose is the most critical step in bioconversion of lignocellulosic biomass to fuels and chemicals. Cellulases and homogenous acids are widely used for cellulose hydrolysis. However, cellulases can only be used at moderate conditions and need longer time to achieve satisfactory cellulose hydrolysis. Homogenous acids can tolerate higher temperature, but they have limitations and issues such as equipment corrosion, recycling and wastewater treatment. To address these issues, heterogeneous solid acids have recently drawn a lot of attention for hydrolyzing cellulose. Traditional solid acids such as sulfonated carbon and resins, however, are not as effective as homogenous acids and cellulases in hydrolyzing cellulose because they have poor access/affinity to cellulose. In this study, a series of porous polymeric solid acids were synthesized for cellulose hydrolysis. These cellulase-mimetic solid acids have hydroxyl, halide, or boronic acid as cellulose-binding group in addition to sulfonic acid as cellulose-hydrolytic group. Their performance in hydrolyzing cellulose and real lignocellulose was evaluated. The properties of the synthesized catalysts such as particle size, surface area, and functional groups and their effect on cellulose hydrolysis were investigated.

(716e) Engineering Strategies to Improve Cellulase Production: Trichoderma reesei as a Pellet

Callow, N., The University of Akron
Ju, L. K., The University of Akron
Kelbly, M., The University of Akron
Originally presented on: 11/7/2013 16:55:00 - 17:20:00

In this study, we demonstrate the use of surfactants and specific nutrient control to regulate the growth and behavior of *Trichoderma reesei* Rut C-30. Although *T. reesei* is the ubiquitous cellulase producing fungus, many practical engineering issues remain. Typical processing conditions utilize a high growth density complicated with high nutrient demand in a viscous filamentous biomass filled fluid phase. This work shows that surfactants can be used to alleviate these processing difficulties by leveraging the inherent cell biology to our advantage. Surfactants have long been used to reduce cellular agglomeration, lyse cells, and improve cellular membrane permeability. In fungal cultures the application the surfactants tends to produce variable, species dependent, responses. In the case of the normally filamentous *T. reesei* Rut C-30, certain surfactants were found to promote growth into a compact pellet formation (diameter 1 ? 4 mm). These physical changes to the morphological structure have an overall positive effect on fermentation, allowing for improved mixing and mass transport through the continuous fluid phase. To improve yields and further decouple cellulase production from the growth phase, the restriction of key nutrients were tested. The specific control of phosphorus was found to successfully halt cell growth without affecting cellulase production. These results indicate that long-term cellulase can be continually produced from a pelleted, stationary phase, *T. reesei* culture.

(770a) A Simplified Route to Prepare Biomass-Derived Terephthalic Acid (TA)

Tong, Z., University of Florida
Wang, F.,
Originally presented on: 11/8/2013 8:30:00 - 8:55:00

Abstract : In response to diminishing oil reserves and increasing costs for petroleum-based products, there is a great need to develop new routes to chemicals intermediates from renewable resources. Here we reported a novel synthesis route to directly transform biomass-derived chemicals (isoprene and acrylic acid) to terephthalic acid (TA) through p-toluic acid (PTA). TA is an important monomer for synthesizing polyesters including polyethylene terephthalate (PET), polypropanediol terephthalate (PPT) or polybutylene terephthalate (PBT), which have broad applications from drink bottles to polyester fibers. The starting chemicals can be produced from biomass through established chemical processes. Bio-based isoprene can be produced by the fermentation of biomass sugars. Acrylic acid can be produced either from dehydration of glucose-based lactic acid or dehydration and oxidation of glycerol (available in large quantity from biodiesel production). In this study, TA was converted from isoprene and acrylic acid through three simple reactions including Diels-Alder reaction, dehydro-aromatization, and oxidation. The simplified process is based on the efficient dehydro-aromatization reaction of cyclohexene carboxylic acid to PTA using a cost-effective catalyst (sulfuric acid). All three reactions were conducted under mild conditions. Finally, pure TA was successfully synthesized with the overall yield of approximately 60% and its structure was confirmed by ^{1}H-NMR (Nuclear Magnetic Resonance). Theoretically, at the first time, we elucidated the dehydro-aromatization reaction mechanism of cyclohexene carboxylic acids using sulfuric acid as the catalyst.

(770b) Co-Fermentation of Glucose and Xylose From Steam Pretreated Wheat Straw With Recombinant S. Cerevisiae Strains

Nielsen, F.,
Zacchi, G., Lund University
Originally presented on: 11/8/2013 8:55:00 - 9:20:00

Advanced bioethanol produced by fermentative conversion of lignocellulosic agricultural residues relies on efficient hydrolysis and fermentation, as well as extensive utilization of available substrates in order to obtain high yields and render the process economically viable. Another important aspect is high ethanol concentration in the fermentation broth, as this reduces the energy demand. Obtaining a high ethanol concentration requires high water insoluble solids (WIS) content in simultaneous saccharification and fermentation (SSF) or in the enzymatic hydrolysis in separate hydrolysis and fermentation (SHF). This usually has an adverse effect on ethanol yield, due to the increased inhibition and poor mass transfer. Opting for either configuration is essentially a trade-off between optimal temperatures and inhibitory glucose concentrations on the one hand (SHF) and sub-optimal temperatures and ethanol-inhibited cellulolysis on the other (SSF). However, the ethanol yield and concentration can also be improved by efficient co-fermentation of glucose and xylose; especially from agricultural residues, such as wheat straw, due to the high amount of hemicelluloses in these materials.

In the present study, SSF and SHF of steam pretreated wheat straw was performed in a variety of process configurations in 2 to 30 L scale. Two different recombinant S. cerevisiae strains, which has the ability to co-ferment glucose and xylose, and two steam-pretreated wheat straw slurries, one pretreated at mild conditions and one pretreated at more sever conditions, were utilized. Substrate feeding and enzyme dosage were varied in order to maximize the ethanol yield. The highest overall ethanol yield achieved, based on the amount of glucose and xylose present in the material fed to the system, was 92% of theoretical for the milder pretreated straw and 84% of theoretical for the more severe pretreated straw. The ethanol concentration was 45.5 and 41.9 g/L, respectively. In both cases the xylose consumption was higher than 90% and the xylitol produced was 1.8 and 3.3 g/L respectively.

(770c) Nutrient Supplementation Key to Inhibitor-Tolerant Yeast Development and Fermentation Performance On Switchgrass Hydrolyzates

Slininger, P. J., National Center for Agricultural Utilization Research, USDA-ARS
Liu, Z. L., National Center for Agricultural Utilization Research, USDA-ARS
Dien, B. S., USDA-ARS NCAUR
Originally presented on: 11/8/2013 9:20:00 - 9:45:00

Inhibitory compounds generated during acid hydrolysis pretreatment of lignocellulosic biomass interfere with subsequent fermentation to ethanol. Tolerant yeast strains have recently been developed by targeted evolution in the presence of biomass hydrolyzate inhibitors. In order to optimize performance, comparative data were collected on both the tolerant and parent strains in experiments testing the impact of nitrogen sources, vitamins, and minerals on ethanol production potential. With respect to optimal ethanol production, a striking difference in the nutritional requirements of parent and adapted strains was observed in both synthetic media and dilute acid hydrolyzates of switchgrass that were supplemented with a variety of commercially available nitrogen sources. Since nutrient availability significantly impacted both ethanol yield and production kinetics, and was strain-dependent, nutrient supplementation requirements were considered in the evaluation and ranking of tolerant isolates as well as in the subsequent optimization and design of hydrolyzate fermentation processes.

(770d) Alkaline Hydrogen Peroxide Pretreatment of Softwood

Alvarez-Vasco, C., Bioproducts Science Engineering Laboratory, Washington State University
Zhang, X., Bioproducts Science Engineering Laboratory, Washington State University

Originally presented on: 11/8/2013 9:45:00 - 10:10:00

Pretreatment of softwood is most challenging largely due to the high lignin content and difficulty in break down lignin compared to hardwood and grass. Acid or SO_2 pre-impregnated steam explosion and organosolv pretreatments have been employed. Both are facing high capital cost and low recovery of hemicellulose. Hydrogen peroxide (HP) is an effective delignification agent and has been widely employed in lignin remove/degradation in paper industry. We also recently found that hemicellulose can be recovered as valuable chemicals from HP pretreatment. This study demonstrated an integrated HP treatment of Douglas fir softwood with the opportunity to simultaneously produce valuable chemicals along with biomass pretreatment in order to maximize the value of both hemicellulose and cellulose.

(770e) Bioconversion of Water Hyacinth (Eichhornia crassipes) At Its Optimum Age Into Bioethanol Via Separate Hydrolysis and Fermentation With Continuous Overliming

Caparanga, A. R., Mapua Institute of Technology
Bathan, R. S.,
Originally presented on: 11/8/2013 10:10:00 - 10:35:00

This study explored the conversion of water hyacinth (*Eichhornia crassipes*) into bioethanol. Prior to bioconversion into ethanol, the plant's optimum age, or age (in days) at which its holocellulose to lignin content was highest, was determined to be 5 days on the average (leaf area ~ 64 cm^2; average stem height ~ 10 cm). At this stage, holocellulose content was highest at 76.3% and lignin content was lowest at 12.3%. Therefore, only five-day old plants were harvested for use in the rest of the experiments. Ethanol was produced from the lignocellulosic materials from the plant by hydrolysis, which breaks the cellulose and hemicellulose polymers to fermentable sugar, followed by fermentation which converts the sugar into ethanol. However, some byproducts such as furan compounds are released during hydrolysis and inhibit the yeast during fermentation. Hydrolysate detoxification using lime ('overliming') was employed. During batch detoxification, sudden addition of lime shows severe degradation of sugar simultaneously with furan compounds. Continuous detoxification was introduced in this study. Its optimum condition was determined to be at pH 12 and dilution rate of 0.667 hr^{-1}. Gradual addition of lime in continuous detoxification could keep the sugar concentration high enough at more than 90% with approximately similar detoxification levels of furan compounds. Using a traditional strain of *S. cerevisiae,* 12.7 g/L ethanol was produced. The ethanol yield per unit biomass was comparable to those reported in other agricultural biomasses: 0.151g/g dry biomass.

(803a) Mixing During High Consistency Enzymatic Hydrolysis - Influence of Raw Material, Viscosity and Particle Size Distribution

Palmqvist, B., Lund University
Kadic, A., Lund University
Liden, G., Lund University
Originally presented on: 11/8/2013 12:30:00 - 12:55:00

In order for commercial scale lignocellulosic bioethanol to be economically feasible, high solid operation throughout the whole process is regarded as a necessity. This will not only reduce the need for high amounts of water and waste water treatment, but also increase ethanol titers after the fermentation step, which will lower energy demands in the following distillation step.

However, since concentrated biomass slurries are associated with very high non-Newtonian viscosities, achieving sufficient mixing during high consistency enzymatic hydrolysis will be a challenging task. Poor mixing can lead to mass- and/or heat transfer problems as well as inhomogeneous enzyme distribution, all with possible yield reduction as an effect. Moreover, we have previously shown that stirrer speed (or more accurately – mixing-power) strongly affects the hydrolysis yield of certain lignocellulosic materials, in particular spruce, where an almost 100 % yield increase was achieved by increasing the stirrer speed from 75 to 500 rpm. On the contrary, other materials such as for example the energy crop Arundo Donax (Giant Reed) seems rather unaffected by mixing power input during enzymatic hydrolysis.

This presentation summarizes what we have learned during the last years about rheology and mixing of high consistency biomass during enzymatic hydrolysis. The fact that different materials respond completely different to mixing power input during hydrolysis will be discussed and seems to be related to the viscosities of the materials. Furthermore, a discussion about liquefaction of biomass versus hydrolysis yield and the effect of particle size distribution on hydrolysis rates will be presented.

(803b) Evolution of Yield Stress During High-Solids Fed-Batch Enzymatic Liquefaction of Cellulose Measured Using a Non-Invasive Real-Time Rheometer

Cardona, M., University of California, Davis
Tozzi, E. J., University of California, Davis
Karuna, N., University of California, Davis
Jeoh, T., University of Califonrnia, Davis
Powell, R. L., University of California-Davis
McCarthy, M. J., University of Calfiornia Davis
Originally presented on: 11/8/2013 12:55:00 - 13:20:00

Enzymatic hydrolysis of cellulosic biomass is a key step of biological routes for production of fuels and chemicals. Economic considerations for large-scale implementation of the process demand operation in the high solids regime. In this regime, the biomass is processed at concentrations of 15% or higher by weight, forming a high viscosity slurry which introduces processing challenges especially in the initial stages of hydrolysis (liquefaction). Rheological measurements with rotational rheometers are challenging in these materials due to rapid changes in rheology that require high time resolution, and the large particle sizes which cause gap effects and settling. We employed a non-invasive real-time rheometer (FlowScan(TM), Aspect Imaging) to characterize the real-time evolution of yield stress of cellulose undergoing enzymatic hydrolysis in a recycle-flow reactor. The hydrolysis was performed in fed-batch mode, with fibers being added at various time points. The suspension exhibited sustained changes in rheology upon initial dispersion of fibers in water (prior to the first addition of enzyme), which was interpreted as hydration or swelling. Each addition of fibers caused a rapid increase followed by a decay in yield stress due to the enzymatic action. The decay of the yield stress was more rapid for the initial loads, becoming slower as the hydrolysis progressed. The influence of fiber loading on the rheology of the slurry undergoing hydrolysis is discussed.

(803c) Maximize Enzymatic Saccharification and Minimize Sugar Degradation in SPORL Pretreatment of Douglas-Fir At a Low Temperature: A Kinetic Approach

Zhang, C., Tianjin University
Houtman, C. J., USDA Forest Service, Forest Products Laboratory
Zhu, J. Y., USDA Forest Service, Forest Products Laboratory
Originally presented on: 11/8/2013 13:20:00 - 13:45:00

Maximize sugar yield and minimize inhibitor formation are often difficult to achieve simultaneously in pretreatment of lignocelluloses. The objective of this work was to evaluate the hemicellulose hydrolysis and degradation models with the goal of predicting performance of both hydrolysis and degradation process and optimizing the pretreatment conditions. Various investigations have developed reaction rate expressions based on Arrhenius and first order rate law. Based on the experiment data, the hemicellulose hydrolysis and degradation kinetic models were developed and the best-fit kinetic parameters were determined. The kinetic models showed that the hemicellulose degradation reaction was more sensitive to pretreatment temperature compared with hydrolysis reaction during pretreatment process. Under the same pretreatment severity expressed by the Combined Hydrolysis Factor (CHF) using SPORL (sulfite pretreatment to overcome recalcitrance of lignocelluloses), a low temperature pretreatment at 165°C for 75min can reduced the inhibitor generation compared with a pretreatment at 180°C for 30min while the substrate enzymatic digestibility was almost identical under these two conditions.

(803d) Comparison of Different Process Configurations Using SSF At High Solid Loadings of Steam-Pretreated Corn Stover

Bondesson, P. M., Lund University
Zacchi, G., Lund University
Galbe, M., Lund University
Originally presented on: 11/8/2013 13:45:00 - 14:10:00

The use of lignocellulosic biomass for ethanol reduces the dependency of fossil fuel and lowers the greenhouse gas emissions. A high ethanol concentration in the broth improves the economics of the ethanol production process. However, when working with high solid loadings, the ethanol yield may decrease due to difficulties to mix, which causes mass transfer limitations. Also, the effect of higher inhibitor concentrations becomes more pronounced. Some of the problems can be overcome by selecting proper configurations for enzymatic hydrolysis and the fermentation process.

In this work washed steam-pretreated fibres from corn stover was used to study the impact of a high solids loading using different process configurations with negligible amounts of inhibiting compounds. A prehydrolysis step prior to SSF or fed-batch SSF was investigated and compared with ordinary SSF using a water-insoluble solids (WIS) loading of 20%. Different variables, such as prehydrolysis time, temperature, feeding time and feeding interval as well as enzyme loading were investigated. The results showed that the inclusion of a prehydrolysis step did not increase the ethanol yield (70-75%) compared with SSF performed batchwise (75%). On the other hand, an improvement was found utilizing fed-batch, which increased the yield (79-82%). Results from this study will be presented.

AUTHOR INDEX

AUTHOR INDEX

AUTHOR INDEX

AUTHOR INDEX